D0783695

A Shepherd's Watch

A SHEPHERD'S WATCH

Through the seasons with one man and his dogs

DAVID KENNARD

headline

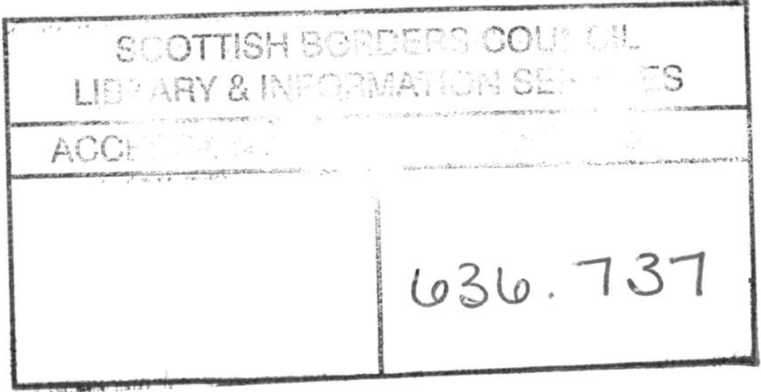

First published in 2004 by

HEADLINE BOOK PUBLISHING

10 9 8 7 6 5 4 3 2 1

Cataloguing in Publication Data is available from the British Library

ISBN 0 7553 1234 1

Typeset in Scala
Designed and typeset by Ben Cracknell Studios
Colour reproduction by Spectrum Colour Ltd. Ipswich
Printed and bound in Italy by Poligrafico Dehoniano

Headline Book Publishing
A division of Hodder Headline
338 Euston Road
London NW1 3BH

www.headline.co.uk
www.hodderheadline.com

For
Clare, Laura, Nick
and Debbie

Gail
Swift
Ernie

CONTENTS

Autumn

CHAPTER ONE

Beginning Again

Mid-autumn, and the telltale sounds of the season were echoing around Borough Farm. The gathering gloom of the past few evenings had been punctuated by the bloodcurdling screech of a barn owl, unseen somewhere in a stand of ash trees where the lower fields meet the woods. Close by, the dense woodland had resonated to the agitated crowing of a pheasant disturbed from his roost, his wings thumping rhythmically as he searched for a new sanctuary. A few hundred yards to the north, the high-pitched cawing of thousands of rooks congregating at dusk had begun reverberating more loudly than ever in the chill October air.

Rookery Wood has been marked on maps of Borough Valley here in North Devon since the nineteenth century. To me the sound of the roosting birds only adds to the mesmerising beauty of the valley at this time of the year. They sweep in each evening from the surrounding farms, in small clusters at first, but then in a jet-black swirl that darkens the sky and fills the air with the distinctive sound of their squabbling, a harsh *kaah kaah*. Each time the flock settles, a new disagreement breaks out, ringing through the birds like an alarm and sending chaotic plumes of them squawking skywards again. The din of their bickering only fades away with nightfall.

The signs were clear that winter would soon be on its way. But it was another sound that confirmed the new shepherding year was about to begin.

I had spent a largely unproductive day struggling to erect a section of fence on a particularly rocky and inaccessible spot, where sheep had taken to breaking from the field into the woods. As the early evening light began to fail, the rooks had started to return home, enacting their raucous ritual once more. Greg and Swift, my oldest sheepdogs, had spent most of the day snoozing under a gorse bush. The crack that disturbed their slumber was unmistakable. With dogs close at hand, I followed the sound and walked in the direction of my small flock of rams, currently resident in the field below the farmhouse.

There's nothing quite like the thwack of two rams fighting on an autumn evening. A warring pair often takes thirty paces back from each other, before running flat out to deliver a butt with the maximum force they can muster. The collision has the same intense ring as two blocks of wood being hammered together. It's a force that would crush a human skull to a pulp. No such danger with a ram. The strength of its skull is quite unbelievable – even though there's not much inside there that needs protecting. Occasionally one will get injured or even killed. But, ninety-nine times out of a hundred, both walk away not even dazed.

Of my eighteen rams sixteen were grazing undisturbed. The other two, however, a black-faced Suffolk ram and a white-faced Texel, were backing purposefully away from each other, heads bowed with malicious intent.

It was no surprise to find the Texel ram was the chief culprit. He was a fine-looking specimen, but since his arrival in the summer had proved particularly stroppy and aggressive. The rivalries had been building up among the rams for some weeks now. The ewes were in season, and the 'boys' knew it.

As it happened, half a dozen of the rams – or 'tups' – were due to go in with the ewes the following day, their introduction timed for lambing to start in twenty-one weeks' time, in early March. But the rams weren't to know 'tupping' was so close, and more and more in the past few days they'd begun sniffing the air to scent the ewes, rolling up their top lips as they did so. The growing tension had spilled over into rivalry, and had culminated in this evening's clash, a classic confrontation over male dominance.

When fighting breaks out like this, the best course of action is to pen all the rams tightly together overnight. Unable to take more than a couple of steps backwards from each other, the rams will be capable of nothing more harmful than a bit of argy-bargy. By the morning any lingering aggression will have been forgotten.

I gave a quiet whistle, and Swift took a sweeping arc to the right of the rams. A word to Greg, and he took the left flank and together they drove the small flock forward towards the yard gate at the top of the field.

Greg and Swift's mutual understanding has developed instinctively during the four years they have worked together as a pair. Very different characters, they form a highly effective partnership. Greg is the brains of the duo, a dog with a laid-back approach to work. Swift doesn't have her partner's intuition and sixth sense, but is a more intense and focused dog who responds to commands with an immediate precision. She is also completely unperturbed by aggression, a particularly useful trait when dealing with rams.

By the time I got to the yard gate the dogs were already holding the sheep there. The Texel, unwilling to accept the authority of the dogs, was once more living up to his reputation. He began repeatedly turning on Swift, his head lowered, making short jabbing movements towards her. At first Swift's reactions were far too quick. His every attempt to land a blow was sidestepped. Swift retaliated only with a warning snap of the teeth around the ram's face. The sparring continued as the sheep moved towards the small, covered pens at the back of the farmyard.

It was as I sent Swift to turn the rams in that the Texel saw its final chance to land a decisive blow. As Swift slipped between him and a wall, ready to make the final turn into the pen, the ram made a headlong charge, catching her across the ribs and pinning her momentarily against the concrete wall behind.

For a fleeting moment I felt despair. With a body weight of around a hundred kilograms and enormous strength particularly in their necks, rams are capable of inflicting serious damage on a dog. Thoughts of broken bones and punctured organs flashed through my mind. For a terrible few seconds I feared I might have lost not just a fine sheepdog but a cherished working companion.

But to my relief, as the ram backed away once more, Swift emerged completely unscathed. Even more amazingly she was utterly unfazed. A couple more snaps of the teeth around the Texel's face and she, with Greg at her side, had pushed the rams into the pen.

The last of the evening light was fading to black. As I crossed the yard heading for home, the screech of the barn owl cut through the quiet again. I scanned the trees at the edge of the outbuildings once more for a glimpse and this time had better luck. The owl's ghostly white silhouette hovered briefly as it hunted near the woods. These birds are rare visitors to Borough Farm. Maybe this one would take up residence, perhaps even staying to breed in the spring.

The screeching grew more distant as I returned to feed the dogs. While Greg and Swift had been at work driving the rams through the yard, my other three sheepdogs, Fern, Gail and Ernie, had struggled to contain their enthusiasm. Ears pricked, standing on their back legs, they had remained shut in their runs, staring intently at the proceedings through the wire mesh, barking excitedly throughout. I took them all for a quick walk up and down the farm lane, then set out their food bowls in their respective runs. They devoured their supper in a minute, then settled down for the night content. 'Good dogs. There'll be plenty of work for you tomorrow,' I reassured them before turning in.

Borough Farm has stood at the southern end of Borough Valley since around 1750. It is little different to the numerous farms that dot the endless hills and valleys of North Devon, except in one respect perhaps. Whereas most of those farms have been built in spots that shelter them from the frequent Atlantic gales, this one stares the elements full in the face.

The Borough Valley is as deep as any in this part of the country. Steep and heavily wooded, its beech, hazel, ash and sycamore would have provided ideal shelter. But for some reason its first residents sited their farmhouse and outbuildings at the head of the valley. Two and a half centuries on, the effects of the pulverising winds are clear to see. On the exposed steep slopes between the farmhouse and the woods a scattering of ash, thorn and beech have been bent crooked. On the edge of the farm a few of the original outbuildings have borne the brunt of the weather and now lean at similarly unlikely angles.

The odd skewed shed apart, though, the farm's original owners would notice few major changes if they revisited the place. Now, as then, the farm's boundaries are marked by immense earth banks, faced with local slate. On Borough Farm alone more than ten miles of these distinctive walls snake their way across the landscape, providing shelter as well as a reminder of an era when the farm would have employed a small army of men capable of building these almost monolithic structures. One of the few noticeable changes to the landscape in the past hundred years is the small pit at the crest of the field, a few yards in front of the farmhouse, where the land suddenly drops steeply to the woods below. The gouge in the earth was apparently caused by a plane, crash-landing at the end of the Second World War.

Of the minor alterations to the old farmhouse itself, the most noticeable would be the double-glazed windows that now look out to the north-west into the valley and the Atlantic coast beyond. When the house was first planned, this seaward side would have been devoid of any natural light. A dark house was considered preferable to a draughty one. Only one window was included, a small square of glass located high up on the wall facing the farmhouse's wide, steep staircase. It remains there today – as does the mystery of why it was added in the first place.

The history of the North Devon coast is rich with tales of smugglers and 'wreckers'. Until the late nineteenth century, the cargo of any ship that came to grief along the coastline became the legal property of the local residents – provided, that was, all the vessel's crew had perished. The impoverished people of small villages like nearby Mortehoe and Lee profited handsomely from the law. Shipwrecked sailors who managed to make it to shore were pushed back into the sea and held under until drowned. On stormy nights, lanterns were lit on lonely cliff paths to imitate the harbour light of Ilfracombe, a few miles to the north-east. If a ship was successfully drawn on to the rocks a shout of 'Ship ashore!' would summon the villagers from their homes. Armed with pitchforks and clubs, they made sure the bounty of the sea became the village's property before daybreak.

Borough Farm is too far from the sea to have been used for drawing in the wreckers' victims. More likely it was used to warn the local lawbreakers of the presence of Customs men. Whatever the truth, there must have been some good reason why the farm was built with such an otherwise useless window, located as it is eight feet from the floor below.

Outside, the most obvious change is the addition of new working buildings. The weatherbeaten original outbuildings encircle the farmhouse on its eastern side. For a modern farm, however, these hay-lofts, loose boxes and calf pens are too small, not to mention too precarious, to be practical. So the bulk of the farm's operations now centre on two large sheds a little further back from the farmhouse.

The first, a large steel-framed sheep shed cut into the side of the hill, was built only a few years ago. Adjoining it is a wooden structure, built in the 1960s of old telegraph poles and ash trunks cut from the woods. The tin roof has disintegrated along its joins, leaving little more than a rusted lacework of metal. The rotten beams underneath don't warrant the expense of a new roof, so I've done my share of running repairs. Every year I decide the shed must be pulled down before it falls down, but somehow it still stands.

In the fields, however, life remains much as it must have been during the early days of Borough Farm. Arable farming has never been a serious prospect here. A few miles inland, the weather tends to be more predictable. Where we are, unforecast rains and heavy sea mists can play havoc with the best-laid harvesting plans. The soil too is unsuitable for growing crops. Even on the best ground at the top of the farm, the topsoil is rarely more than six inches deep. Any attempt to plough the ground would succeed only in turning up shillet – and doing serious damage to the machinery. As you move down closer to the sea, there is even less topsoil. At the bottom of the farm, the rocky outcrops protrude in long ridges. So while dairy farming dominates a little further inland, here on the coastal strip sheep farming has always been in the ascendancy.

Of all the changes the farm has seen in the past 250 years, the most significant are the commercial realities that have recently hit sheep farming. At the turn of the century, Borough Farm grazed 270 ewes and forty head of cattle. Its owner employed three full-time workers, and would have been considered a wealthy farmer. Today, I keep a little over eight hundred ewes at Borough Farm and on the 260 acres I rent from the National Trust at nearby Morte Point. In a good year, between the two flocks, they produce 1,200 or so lambs. Eight hundred ewes make for a full-time job, but the days are gone when this size of flock can – on its own – provide a living for a family like mine: my wife Debbie and me, and our three young children, Clare, Laura and Nick. Sheep farming has always been prone to the problems of weather, disease and unpredictable market prices, all of which are largely outside the control of the farmer. But such are the uncertainties today that a major problem with the ewes or the coming year's lambs could push us over the edge financially. We are already reliant on Debbie's income from her job at a local nursing home. A bad year, and I would have to consider cutting back the sheep and getting a 'proper' job myself.

There is no doubt in my mind that in a generation or two – maybe even less – sheep farming and shepherding will no longer be a viable way of life. I certainly couldn't encourage any of my children to take over from me, even if they wanted to. It's a sobering, saddening thought, but I do sometimes wonder whether I represent one of the last generations of shepherds.

CHAPTER TWO

The Battle of the Ram

I emerged from the house and walked across the yard to be greeted by five barking, whining, yelping dogs. Greg, Swift, Gail, Fern and Ernie were peering through the wire of their kennels in my direction, their heads wobbling from side to side as if to counterbalance the wagging of their tails at the opposite ends. The decibel level went up another notch as I set the whole pack free, then made the whooshing noise that was my signal for them to career up the drive before returning to face me, awaiting further instruction.

It wasn't difficult to spot the pack's pecking order. When at play, each dog chases its immediate senior, usually biting playfully at the scruff of its neck along the way. So this morning, as usual, seven-year-old Greg, the undisputed top dog in my kennel, was being pursued by Swift, two years his junior and the top bitch. She, in turn, was being chased by eighteen-month-old Fern, who was being pursued by Ernie, at eight months the youngest dog. The odd one out was Gail, Greg and Swift's twenty-two-month-old daughter, who for unknown reasons has always seemed aloof from the established hierarchy. Gail ambles along in a world of her own, seldom leaving my side. Perhaps it is to do with the fact that she is the victim of the only friction in the kennel. Fern and Gail are bitches of a very similar age, and have never got on. In my presence there is rarely any problem, but I can never leave them alone together.

It was Ernie, though, who required my attention this morning. Even at eight months old, he is potentially the best sheepdog I've ever had. 'Potentially' is the key word. Ernie is totally obsessed by sheep and would spend every waking moment doing what comes instinctively to him. But until he's been properly trained he's an absolute liability around the flock.

This morning he quickly lost interest in the morning's run. Instead, guided by smell or sound, I'm not sure which, he had been drawn to the rams that had been left overnight in the sheep pens. His ears were pricked and it was clear he was desperate to work them. He wanted to take them somewhere, anywhere – he really wasn't bothered.

I called him repeatedly as he made a beeline for the pen. Momentarily he took a few paces back towards me, but within seconds he turned again towards the rams. The increasing frustration in my voice was being completely ignored, so I resolved to leash him. I was unravelling a piece of baler twine from my pocket when I realised I was already too late. As I walked around the corner to the pens, I saw Ernie leap the gate into the rams' midst. The suddenness of his arrival set the animals backing away into a packed huddle in the corner of the pen. With the rams on the retreat, he prepared to press home his advantage.

In the past few weeks, Ernie had developed a tendency to dive in and grip on to sheep. It is not uncommon for a puppy to pull wool – the sheepdog's instinct to work is a refinement of the instinct to hunt. But Ernie had a very broad head, and a strong muscular jaw, and on one or two occasions had gripped on to leg rather than wool.

Although it is a major flaw in a dog, I felt sure that it was down to Ernie's over-exuberance and that I could train it out of him in time. Fortunately so far he hadn't caused any damage, but it was a bad habit – one that I must prevent him from displaying again. I ran over, climbed in and intercepted him just as he attempted to dive behind a somewhat bemused Suffolk ram.

I have always liked a puppy who was keen to work. Ernie took that keenness to a new level, however, and it was becoming impossible to let him run free anywhere on the farm. Reluctantly I decided that his help this morning was more than I could cope with. I had no choice but to cut short his exercise and return him to his run.

An old shepherd friend of mine once told me about the 'advice' he had given a novice sheep-keeper. 'If you want a decent night's sleep during lambing time,' the old-timer had said sagely, 'take the rams away from the ewes at night.' It was, of course, a joke, but one rooted loosely in fact. The gestation period of ewes isn't quite predictable to a precise hour, but it can be anticipated with amazing accuracy, usually at 147 days. There is slight variation between different breeds, but in general few ewes will lamb outside the 145- to 150-day range.

The great advantage of this is that during the breeding season – or 'tupping' – a shepherd can plan the dates of the following spring's lambing pretty precisely. Ewes have a cycle of seventeen days, during which time they are fertile for three days, so lambing should start twenty-one weeks after the rams are introduced. Barring a few stragglers, it should be over three weeks after that.

In choosing the precise timing of lambing, there are a host of factors to be weighed up. The most obvious are the weather and the supply of fresh new grass the farm is likely to have in the spring. There are wide variations in the time lambing begins, not just across the country but even in North Devon. Because Borough Farm is in an exposed location, it isn't what is termed an 'early' farm, so I lamb the bulk of my flock towards the end of March. Yet only fifteen miles away, on the higher, wetter ground of Exmoor, lambing is generally left three weeks later than this. In the hills of Scotland and Wales, it can be May before shepherds feel confident of the spring weather they need.

The lambing calendar isn't entirely the shepherd's to control, of course. The ewes themselves determine to some extent when lambing is possible. Their 'heat' cycle doesn't commence until the hours of daylight decrease in the early autumn, and some breeds naturally start their cycle earlier than others. The breed I keep, North Country Mules, will begin to 'take the ram' only from mid-September.

The final consideration is the lamb market. The fastest-growing lambs can be ready for sale at ten or twelve weeks of age, but most will not leave the farm until they are between four months and a year old. Inevitably the laws of supply and demand come into play. With most flocks in the country lambing from late March onwards, the lamb market can be flooded by the end of June, sending the price crashing. It's not until the end of September that prices tend to rise again.

So, given that the summer is financially the wrong time to be selling the bulk of the lamb crop, my strategy is to lamb a hundred ewes at the beginning of March. It means I have just enough shed space to be able to keep the lambs inside if the weather is really bad. When the weather allows me to turn the lambs out, I can feed them in the fields where – with luck – they will be ready for market before the prices start to dip. The added bonus then is that I will have a hundred ewes who require far less grass without their lambs. And, of course, we will also have some money in the bank.

All this means that the success of the following year is largely determined in the autumn. And the most crucial factor at this time of the year is ensuring both the rams and the ewes that are going to be put to them are in the best possible condition. Over the past six weeks, each of the farm's eight hundred ewes had been assessed individually. The key is to maintain a flock that is capable of producing and rearing viable lambs, so the main task had been to sort the ewes according to their ability to do this.

Mastitis is a common problem – a ewe affected in one season will rarely produce the milk required to feed her lambs the following year. So I had begun by checking the flock's udders for lumps. In addition the ewes' feet were checked, any overgrown hooves were trimmed and the all-too-common foot infections treated.

Most sheep tend to begin losing their teeth at around five years of age, but the North Country Mules I keep tend to lose them earlier than other breeds. On the ample summer grazing this may not be a problem but, as the summer grass disappears, such sheep often struggle to maintain their condition. Those with bad teeth or who seemed too thin to survive another winter were sold off.

With the core ewe flock chosen, every animal was then treated for worms. Finally, I 'docked' them, shearing the wool from their back ends, to prevent them becoming soiled over the winter.

At the end of this process, the flock was divided into three groups, what you might call the fit, the fat and the 'fin'. Each group had then been taken to suitable pasture so that by the time they met the ram, each individual was going to be in the best possible condition, to conceive the maximum number of lambs.

If the ewes need to be prepared for the marathon that is a long winter of carrying lambs, the rams need to be readied for the sprint that is their busy month of the year.

There's no question that a ram's life is a strange and, at times, a stressful one. As a proportion of their body weight, rams have the biggest testicles of any animal. They need them because they have a Herculean task ahead of them during the tupping season. An average ram will serve anything between forty and a hundred ewes, but may well serve each one half a dozen times. Usually his work is spread out over the full seventeen days of the ewes' cycle, but it can be more intense. The introduction of a vasectomised, or 'teaser', ram into the ewe flock early in the year has the effect of getting the flock to start synchronising their cycles. By tupping, the entire flock can be ovulating during the same three-day period. On one occasion I put seven rams in with 150 ewes, which had been run with a 'teaser' ram. It was only eighteen hours later that I next inspected the flock, but 118 of the ewes had been served. As I had arrived in the field, seven rather weary and hungry-looking rams were dragging along towards the back of the flock. Their expressions read as if to say: 'Let me tell you about the night I've just had.' It was an admirable feat of endurance.

When 'tupping' is finished, rams have eleven months off without a sniff of the action. But this intense month of activity takes its toll. The rams can lose up to thirty per cent in weight during the tupping period. By the end of November, my fine-looking male flock will be a shadow of their former selves, so they have been fed heavily for the last six weeks in preparation.

With the over-exuberant Ernie safely inside his kennel, I started to prepare the rams to join with the ewes. The first task this morning was to fit the rams' harnesses, an important tool for the shepherd. The leather harnesses carry a coloured crayon that sits across the ram's chest and leaves a mark on the ewe's rump when she is served. As well as identifying which ewe has been served, it is also a long-lasting record of when this happened. For instance, by using a blue crayon for the first ten days of tupping, and a red for the second ten, the flock can be divided into two, those to lamb earlier and those to lamb later. Using other colours, it's also possible to identify the breed of the ram by which the ewe has been served.

I called Swift and Gail into the pen in which the rams had spent the night. Their enclosure formed part of the handling pens, a series of gates and pens designed to help administer the various treatments required over the year. The rams' pen was linked to the treatment pen where I was now going to fit the harnesses.

Swift quickly brushed past the rams to take up a position to force them into the treatment pen. Gail was a little reluctant to force her way past. She sensed the aggressive nature of the rams and made two or three abortive attempts to pass them before finally seeing a gap and making a dash to join Swift.

The rams possess a fearsome strength in their necks, occasionally matched by a cussed temperament. I backed each one into the corner of the pen, and struggled as each in turn lowered his head obstinately to the ground, trying to resist my attempts to raise his chin in order to fit the harness. The same old Texel ram caused the most problems, wildly flicking his head at my hand as I attempted to arrest his chin. Before I had him restrained, he had delivered a short jabbing butt that narrowly missed my knees. It was only by backing him into the corner, pressing his head between my thighs and covering his eyes, that I pacified him.

At the opposite end of the treatment pen is a small gate, big enough for only one sheep at a time. The treatment pen can be exited only through this gate, which opens into a single-file 'race' – a narrow passage twelve feet long. Sheep have a natural desire to follow the sheep in front, which means that once the first of them has entered the race a constant stream of others follow on behind.

I was going to put out six rams today – the rest could wait for three weeks. With half a dozen rams successfully harnessed, I opened the small gate to the race, and the rams jostled to be first in. At the opposite end of the race is the 'shedding' gate, which switches from side to side, allowing the flock to be diverted in two directions. I took the handle on the gate and changed the gate in front of three Suffolk rams, which I'd decided would be the first to join with the ewes. As the last ram came through, followed closely by Gail and Swift, I switched the gate in front of the dogs, who obligingly followed into the pen holding the three Suffolks. The other fifteen rams were left in another pen.

For the three Suffolks it was only a short walk a little way up the farm drive to a flock of ninety ewes with which they were to be joined. There was a tangible air of excitement in the field as the rams arrived to be quickly surrounded by a group of amorously sniffing ewes. I left them – and Mother Nature – to it.

Back at the pens, I collected the twelve rams that weren't needed today and took them back to the paddock with Gail and Swift. I then returned to the sheds for the remaining three rams, a trio that included Swift's old friend the problem Texel. They were going to a field at the other side of the farmhouse, where I had another hundred or so ewes waiting.

Immediately I opened the gate out of the pens, the sparring started again. The Texel, apparently objecting violently to being bossed by what he considered inferior animals, was far more intent on landing a blow on Swift than keeping up with his colleagues. Swift moved her head quickly from side to side just in front of the ram's. Each act of aggression was met by a snap of Swift's teeth, which made the ram retreat a step or two. Each time he backed off Swift advanced, denying him the chance to take a run up and deliver a more forceful blow. Gail tucked herself in a foot or so behind Swift, slightly unsure of herself in the presence of such an aggressive customer.

Progress, already slow as we made our way across the yard, took a turn for the worse as we headed left towards the paddock behind the farmhouse. The Texel ram saw his chance to escape, breaking off from his two colleagues and heading in the direction of the house. Swift went to turn him back once more, but before she could get to the right side of the ram he had reached the French windows. For a second the ram stood there, looking at his reflection. Presumably mistaking it for a rival, he then proceeded to run headlong into the glass.

The toughened glass shattered with a loud crack, but thankfully stayed in place. As if to prove he was completely unharmed, and showing no remorse, the Texel immediately resumed his battle with Swift. Gail took another few steps back, increasingly alarmed by the events.

I decided I'd had enough of this ram by now. In the space of twenty-four hours he'd tried to assassinate one of my top dogs, almost kneecapped me as I'd tried to harness him and now, finally, tried to destroy my home. I drove him the last few yards myself, nudging him from behind and waving my crook firmly across his face whenever he made any further attempts to turn away from the direction he was supposed to be heading.

As we finally shut the gate behind the three rams, Gail and Swift turned back out of the field. I patted my leg and Swift jumped up, resting her front legs on me. I rubbed her nose affectionately. 'Just as well they're not all like that, Swiffie.' She looked up. I'm sure she knew exactly what I meant.

I headed back into the house for a cup of tea. Swift and Gail slipped in the door behind me, but fortunately I spotted the green slurry colour of their feet and bellies before Debbie did, and quickly ushered them back out the door, wiping their green tracks from the kitchen floor with my socks as they went. I was still feeling frustrated by the antics of the rogue ram, but any slight hopes I might have had of finding any sympathy in the house quickly dissipated.

Nick, nearly three, was – as ever – lying on the living-room floor making loud 'brrrrm, brrrrm' noises with his toy tractor. Clare, nine, and Laura, six, had just got back from school and were hanging up their coats and bags in the hallway.

'What happened to the window, Dad?' Clare asked, looking mildly shocked at the shattered glass.

'That stupid Texel ram stuck his head through it,' I told her. Clare is a real country-loving girl. From an early age, she's always been keen on helping on the farm, and will often give me a hand for half an hour before going to school. She has her own rabbit and guinea pig, as well as a personal sheep that she sprays with her name at lambing time. Her greatest passion of all is horse riding, and she has lessons each Saturday at the stables in the village. She is constantly nagging for a pony of her own.

My only concern is that she's turning into something of an animal rights activist.

'Well, it was probably your fault,' she said. 'You probably upset the poor thing. You should have been more careful.'

'"Poor thing"!' I protested. 'That ram's spent the last two days trying to kill Swift.'

Until that moment Laura had been getting on with things, minding her own business. Laura is quieter than her big sister and tends to do what Clare tells her, albeit reluctantly at times. She's not quite as keen on the outdoor life as Clare, but she adores the dogs. Laura had been very worried when I'd told her about the ram's attack on Swift the evening before. Swift has a very soft side to her nature – when there's no work to be done she likes nothing better than to be stroked and petted by the children, Laura in particular. The prospect of Swift being hurt horrified her.

'Is Swiffie all right?' she said, a look of real concern on her face.

'She's fine, Laura,' I consoled her. 'She knows how to look after herself.'

'I'd better go and check on her anyway,' she said. 'Where is she now?'

'Just outside,' I said.

'Can I bring her in?' she asked, going to the cupboard under the sink to collect a chew for the hero of the hour.

'Probably better not,' I said, looking at the green smudge still stretching towards the back door.

As I went to inspect the damage to the French window, Debbie appeared from upstairs. This wasn't the first time a sheep had wandered too close to the house. She'd been gently going on at me for some time about a fence to keep them from occasional forays into her vegetable garden. Debbie's a laid-back person, not the kind to indulge in histrionics or 'I told you so'-style recriminations. Having herself grown up in the countryside with a small flock of sheep and goats in the back garden, she knows how downright obstinate sheep can be. She simply gave me a look, which said it all.

If I had any doubt about her feelings, Nick was soon confirming them. It turned out that he had been playing with his tractors close to the French windows and had witnessed the ram's display of bad manners. Like most boys of his age, he tended to keep his sentences brief and to the point.

'Daddy, why did the sheep crash the window?' he inquired, pointing to the scene of the incident. He then looked at me with a very serious face indeed. 'Mummy's cross,' he said. Just as well she hadn't seen the mess on the kitchen floor, I thought.

CHAPTER THREE

Greg

Behind the mesh grill of my ageing Land Rover, Fern was moving restlessly from side to side, emitting a mixture of high-pitched whines and suppressed yelps as we bumped up the lane and out on to the road. Alongside her, Gail and Swift could not have presented a more contrasting sight. Mother and daughter were sitting calmly, seemingly without a care in the world. It was Greg who was causing me the greatest problems.

Greg is undoubtedly the top dog in the pack, a position he maintains by physically dominating the younger dogs. So whenever we go out in the Land Rover he presides over what he considers to be the position of importance, sitting directly behind me. As if this isn't enough, every few minutes he reaches his cold wet paw through the mesh, and places it down the back of my neck – just to remind me he's there. Today my requests – 'Greg don't DO that' – had proved a complete waste of breath. Drips of muddy water were running down my back, and the lines left by his claws were feeling a little sore on my neck. He was only trying to be friendly, I told myself.

As I did three or four times a week, I was making a short trip to a farm near the village of West Down. The farm was providing a temporary home to the two hundred lambs I was keeping there during the winter months.

In the run-up to tupping, it had been time to offload some of the stock at Borough Farm. About half of last year's 1,200 lambs had been sold, but there were still just under seven hundred on the farm, now about six months of age. With the colder weather and the decreasing hours of daylight, the grass was going off, disappearing quickly. The nutritional value of the grass was going with it, which meant the lambs were not going to 'finish'.

'Finish' is the term used to describe the amount of flesh covering on an animal, something which is assessed by feeling the back, ribs and tail. Lambs need a sufficient level of nutrition to achieve enough finish to be sold at market. They weren't going to get that on a sheep farm that saw sheep every day of the year, and where any fresh grass was now being kept for the ewes whose need was greater.

By a happy coincidence, it's also around this time every year that the dairy farmers are finding that the weight of their cows is 'poaching' or churning up the ground. As they bring their cows in by mid to late October, or earlier on a wet year, so the sheep come in useful. If it isn't eaten, their grass dies off and inhibits the growth of the fresh grass when it comes through the following year, so many of them take the option to have sheep in to clear off the last of it.

I have three local dairy farmers who will take in lambs for eight or ten weeks over the winter. The effect it has on the lambs is quite remarkable – they finish in no time at all. It does, however, mean a lot of time spent travelling, looking after the lambs. Dairy farms tend not to be fenced for sheep, a single strand of barbed wire being sufficient to keep in cows. A lot of work is created in putting up electric fences and checking the lambs haven't broken out.

I had got the hint from Greg that he wanted to come out to work first. So when we arrived at the fields where the lambs were grazing I opened the back door with a firm: 'Just Greg.' He was already out before any of the others had a chance to try and squeeze past him.

West Down is typical of one of those picture-book Devon villages, surrounded by green rolling hills, gently sloping down lightly wooded valleys. From the field where the lambs grazed there is a striking view, far across the North Devon coastline to Bideford Bay, nearly twenty miles away. In the valley below, at the far end of the fields, lie some caves, formerly copper or zinc mines. They were now home to a rare colony of horseshoe bats, which could be seen departing the cave in small clusters at sunset.

This particular flock of lambs had the run of three fields, each of around eight acres. About half the flock was grazing in the first of the fields. As the main priority of the visit was to count them, I sent Greg off to his right to return them to me. The lambs started to run towards the far gate, startled by the sudden intrusion. Instead of continuing on the logical course, however, Greg veered off sharply and stopped, his attention drawn towards the furthermost edge of the pasture where the banks were overgrown with thick bramble.

I had bought Greg seven years earlier, as a rather large, lolloping, five-month-old pup. He had quickly proved an exceptional dog, with a natural ability to work, and had taken little training. His great strengths from the beginning were his brains and vision. Time and again, Greg had displayed an ability to think, to reason his way around situations.

During the course of a working life spent together, a shepherd and his dogs develop a deep understanding of each other's behaviour. Small signals can speak volumes. So I knew that if Greg had stopped there was likely to be a good cause. He probably had something to tell me.

As I watched he jumped on to the bank, his ears half pricked, looking intently at something on the other side. A second later he jumped down and out of my sight. He appeared again within moments. This time he deliberately looked back at me, then down intently once more at the other side of the bank. His actions were purely instinctive, something I couldn't possibly have trained a dog to perform. This was Greg's way of communicating a situation he knew to be a problem. All I had done during our years together was learn to understand what he was telling me at times like this.

I strode across the field towards him, confident that I knew what the problem was. Sure enough, on the far side of the bank was a lamb, its thick woollen coat entwined in a mass of brambles. This was the most common problem with putting lambs out on to dairy farms. In summer, brambles don't stick to cows' tough hides so cause little problem to dairy farmers. By autumn, and the arrival of the lambs, however, the brambles have grown thick and are particularly vicious. Once, after a particularly rough night had driven the lambs to look for shelter, I had spent the following morning pulling more than forty of them free from the brambles.

This particular lamb wasn't helping the situation by constantly twisting and turning. Rather than releasing him, each movement was transforming the multitude of bramble stems into a thicker rope and an even more constricting straitjacket.

I was always impressed by the way Greg's mind worked. He had heard a movement, or perhaps a noise, from something that was out of his own range of vision. Jumping on the wall to investigate, he had first attempted to free the lamb by gripping the wool with his teeth and pulling. Having had no success, he had jumped back on to the wall, looking towards me and then towards the lamb – his way of communicating that he needed my help.

After I spent a few minutes cutting bramble stems with my pocket knife, the lamb was on its way. Greg momentarily looked up at me again and gave half a wag of his tail. I gave him another command and he was off, another two hundred yards, instinctively turning through the gateway into which the rest of the lambs had disappeared.

As well as brambles, keeping sheep on a dairy farm during the winter presents a second problem. While a well-fed cow will remain in a single field contentedly chewing the cud, sheep have a very different mentality. The saying 'The grass always looks greener on the other side of the fence' was, I feel sure, coined by a shepherd with lambs 'at keep' during the winter. By the time I entered the second field there was no sign of the rest of the flock. Nor was there a whiff of them in the third field that had been allocated to them.

Keeping the peace with my temporary landlords is one of the priorities for the winter. The one thing guaranteed to strain my relations with a farmer is him receiving an irate phone call from his neighbours about 'breaking sheep'. 'Where the hell have they got to?' I muttered to myself. Once again, however, Greg was several steps ahead of me.

Two hundred yards away I saw him disappearing over a low point in the wall at the far end of the field. A few minutes later I saw the first of the partial flock of lambs, clambering back over the wall. It was only when the last of them were back in the correct field that Greg reappeared.

Once more I was left scratching my head at his incredible instinct. On closer inspection of the wall, I saw that there was a muddy patch where the lambs had been climbing over. Greg had never been in these fields before, yet somehow he had seen the mud on the wall then worked out what had happened. Perhaps he had heard them, perhaps they'd bleated, but somehow he had sensed what the lambs had done. It was another reminder of the extra level of intelligence that gave him an advantage over not just the rest of my own dogs, but most of the sheepdogs I've ever come across.

I ran the lambs slowly back through the gateway into the field where the rest of the flock were grazing, counting them as they went. Satisfied they were all gathered up, I turned back to the Land Rover once more. Greg ran ahead sniffing in the long grass, turning playfully towards me now and then. The look in his eye and the wag of his tail were those of a dog rather pleased with himself. I wondered whether he knew how highly I thought of him.

As I reached the final week of October, the tupping was going well. In the space of five days or so, the first batch of six rams had served the hundred ewes due to be lambed early. I'd then removed the rams for a fortnight before uniting a dozen of them with the Borough Farm ewes last week. Each morning since then I'd counted around thirty more ewes displaying a brightly coloured rump. Half of the Borough Farm's six hundred ewes had now been successfully served. It was now time for the last batch of six rams to join with the 250-strong flock out on Morte Point.

North Devon is blessed with some of the finest scenery in the country, but there can't be a more striking location than Morte Point. At the northern end of Woolacombe Bay, the peninsula looks out over Lundy Island, thirteen miles off the coast. Beyond, the Welsh coastline and even the Black Mountains are visible on clear days.

The Point's westerly tip is dominated by a stark spine of rock, layered slate that breaks the skyline like the back of an armadillo, its tail extending, largely underwater, half a mile out to sea as far as the infamous Morte Rock. The peninsula itself is a rare coastal heath and provides a habitat for a thriving population of small birds – stonechats, linnets and warblers among the thick scrub of the rocky slopes, pipits and wheatears on the rough grassland, oystercatchers and rock pipits on the cliffs and coves. Closer to the sea, cormorants, herring, black-backed gulls and the occasional peregrine falcon patrol the Point's feeding ground. For me, the bird life only adds to the exhilarating, endless beauty of the place.

For all its rugged grandeur, however, the Point is an exposed place. While the Borough Farm flock graze on 170 acres of reasonably sheltered ground, the sheep that graze on these 260 acres of coastal and cliff ground spend their lives in a far harsher and more demanding environment. As a result, the harder weather and poorer grazing here demand a slightly different approach to sheep farming. So the flock is based around the Romney, the native breed of Romney Marsh in Kent, now the most widespread sheep in the world. The Romney is a tough, durable sheep, more than capable of coping with the conditions here. I cross some of the ewes with Romney rams, to keep the breed pure. The remainder are crossed with Texel rams for the quality of the meat, or with the Lleyn, another hardy breed, this time from Wales, whose offspring tend to produce more lambs than the Romney.

There's a local saying, 'If you can see Lundy, it's going to rain; if you can't see Lundy – it's raining!' Today I couldn't even raise my head in the direction of Lundy, such was the volume of rain being driven in by a particularly violent squall that had been blown in from the Atlantic. I opened the back door of the large covered trailer, keeping the hood of my jacket down over my eyes to protect them from the stinging rain. I dropped the tail board and joined the rams inside while I fitted their harnesses.

As I keep ewe-lambs back in this flock to become replacement breeding ewes, it is important to avoid in-breeding. I do this by using a simple identification system. During tupping I fit the Romney rams with a distinctive red marker crayon, the Lleyns with a blue, and the Texels with a green. These should still be visible come lambing time. In the spring, each time a ewe-lamb is born to a mother who has been served by a Romney ram, I take a small ear-notch from the front left ear of the lamb. When the lamb comes to breed this means I will know it must be served by one of the other two breeds. Similarly, the ewe-lambs born to those ewes served by a Texel or Lleyn ram will be given an ear-notch in the back of the left ear, or front right respectively. When they come to breed, they will be served by Romney rams.

The rain rattled on the tin roof, and every now and again a gust of wind rocked the trailer slightly. Inside the air felt warm and damp. For a moment it was reminiscent of the wet camping holidays of my childhood. Unsurprisingly, for once the rams showed no inclination to slip out of the door. I almost felt guilty at forcing the first two Romney rams from the trailer. They held their heads low against the storm and bolted instinctively to the cover of a nearby gorse bush.

With a good waterproof jacket and leggings, I have no problem with wet conditions. I even quite enjoy it. Greg and Swift accompanied me, eyes half closed against the rain but otherwise unaffected by the pounding winds.

Morte Point was given to the National Trust in the 1930s by Lady Chichester of Arlington. The bequest is commemorated by the large steel gates that mark the entrance to the Point, near the village's cemetery. The beauty of the headland and the surrounding coastline attracts many thousands of visitors during the summer months. The numbers dwindle through the autumn and winter, and it's largely people from the village that I meet out on my regular inspection rounds. Today, however, there wasn't a soul to be seen. The Point was all mine. The solitude of shepherding is one of the great appeals of the job.

We guided the two rams over the hill and down a small path that winds its way through the rich brown of the dying bracken. The violent squall at last abated with the same suddenness with which it had arrived. The threatening dark grey clouds began to clear away inland. Out to sea, a patch of silver light on the water promised a little brightness to come.

As we approached, down the steep bank, the ewe flock stirred from their shelter, tucked in tight to the leeward side of a patch of gorse and blackthorn. As they emerged into the winds, each of the ewes shook herself, expelling a great shower of collected rain. Scenting the arrival of their menfolk, the flock gathered excitedly around the rams. It wasn't long, however, before the mood had settled and most of the flock had drifted away, leaving half a dozen ewes, obviously in season, clustered around the two rams. I joined the exodus and again left nature to take its course.

With Greg and Swift, I took the longer route back, along the cliff path. The patch of sunlight had spread in from the sea, breathing new life into the bracken, now a deep glowing gold. The yellow of one of the odd, autumn-flowering gorse bushes radiated an almost fluorescent glow.

Gazing out to sea, I caught a glimpse of a white flash dropping into the grey-green waters. Straining, I could make out first one, then a small flock of gannets, dropping twenty feet from the air on to a shoal of fish, possibly mackerel. Five seconds or so after they disappeared, each bird would reappear. Some would take to the air and dive again, others floated on the sea, presumably the new owners of a successful catch.

I turned back for the Land Rover, removing my jacket to let out some of the dampness. At the landward end of the peninsula, I could make out the small cluster of ewes and rams in a valley beneath me. Even from that distance it was clear that the rams had set about their work with all the vigour you'd expect of anyone who hadn't worked for a year. The final countdown to lambing was under way: 147 days to go. The cycle of life was in motion once more.

CHAPTER FOUR

Windy Cove

Morte Point is never short of natural spectacle and, with the Atlantic winds driving in from the north-west, today was no exception. On the north side of the Point, huge waves were piling into the spine of the headland, sending geysers of water shooting high into the air. In weather like this it wasn't difficult to imagine why this stretch of coastline has been infamous among mariners over the centuries. The local heritage centre carries the names of seventy-seven vessels which, according to Lloyd's Register, perished here. Thanks in part to the efforts of the wreckers, many more almost certainly foundered, their losses unrecorded.

I'd cast my eye over the ewes Greg and Swift had turned in from the comparatively sheltered southern edge of the Point and was watching some oystercatchers sheltering on an earthy ledge below the lip of the cliff, when I was aware of someone approaching me.

'Are you the shepherd?' the walker asked, his expression suggesting he'd regretted the question even before he asked it. With a crook in my hand and three collies at my side, I could hardly have been anything else. 'There's a sheep stuck on a ledge over there.'

I was a little sceptical – and not just because the cliff to which he was pointing was so sheer I couldn't think of a place where a sheep could possibly become stuck. During the course of a year I get frequent reports from holidaymakers of cases of sheep suffering from a variety of ailments. Some claim they have seen a dead sheep when in reality they've only come across a ewe enjoying a sleep in the sun. I've even had people stop me and tell me they've heard one of them coughing. In this case, however, the walker wasn't exaggerating.

At a spot called Windy Cove, a ewe had got herself into a position that was precarious – to put it mildly. Somehow, I didn't know how, she had got herself stranded on a ledge, roughly ten feet beneath the edge of the cliff. The ledge itself was hidden until you stood directly over the top of it, which was apparently what the walker had been doing while watching the storm unfold. From any other angle both ledge and sheep were completely invisible. As I approached from the edge of the Point, I saw the narrow terrace where the ewe was stranded was overhung from above. Even if she'd wanted to, she couldn't get back up the way she'd come down. Forty feet below her, the relentless waves were battering the jagged rocks. A slip in the wrong direction and she would be heading down into them.

With two miles of unfenced cliff, sheep that have got stuck on ledges through their own stupidity, or chased on to the rocks by a stray dog, are a fairly regular occurrence out on Morte Point. Generally my philosophy tends to be based on something a vet once told me while treating a friend's horse: 'Physician, do no harm.' In ninety-nine out of a hundred cases, a sheep will find its own way back to higher ground. Trying to retrieve it can cause panic, which can lead to real problems.

This was a more complicated situation, however – and not just because the sheep was in such an awkward spot that she had no hope of extricating herself.

Close to a dozen curious onlookers had seen what was happening and had positioned themselves in front-row locations along the cliff path in anticipation of some dramatic action. As if the unwelcome audience wasn't enough of a nuisance, another one was tugging at the other end of the length of baler twine I had in my hand: Ernie.

These past few weeks, Ernie's daily exercise had tended to be limited to a training session or two, and a run around a field void of sheep. I had a soft spot for Ernie, however, and today I'd succumbed to the urge to bring him with me. The idea had been that I'd give him some exercise and a little gentle schooling on the lead, but instead Sod's Law had applied itself. With a ewe on the edge of a cliff and a gallery forming to watch me rescue her, Ernie had become a huge liability. The thought of his assistance at the moment was – well, unthinkable.

He had already spotted the ewe and was straining at the string, whining as he pulled with all his strength in the direction of the cliffs. Securing him to a nearby gorse root was the obvious answer but, on previous occasions when I had attempted to restrain him with string or rope, he had merely bitten through it, with remarkable speed. I resorted to tying him up, and scanned the faces of the onlookers for help. Fortunately I recognised Tom, a young lad from the village, who was out walking with his parents. I asked him whether he'd mind keeping an eye on Ernie while I dealt with the ewe.

'Whatever you do, don't let him chew through his lead,' I told him.

'OK,' Tom replied, looking rather pleased to be involved and be given a job of such importance.

'Sheep sense' is an indefinable commodity, something that those of us who spend years in the presence of the ovine species develop. Put simply, it's an ability to know what the instinctive behaviour of a sheep will be – often well in advance of the animal itself. As Swift and I slowly started to climb down towards the ledge, the message I was getting back from this ewe was 'One step nearer and I'll jump'.

The only option was to get down to the ewe and try to physically lift her up. Inevitably this meant that I needed to get to the lower side of her, and use Swift to block her only possible escape route, a thin ridge running further down the edge of the cliff. I wasn't going to do anything that was likely to risk life and limb. At the end of the day, it wasn't worth it for a sheep. This was 'doable' – but it was at the limit of what was actually sensible.

Swift is fantastic in these sorts of situation. She's tremendously manoeuvrable, and has never had any fear of the cliffs. Dogs tend not to suffer from vertigo, but fortunately do appear to understand the potential danger of heights. The biggest risk to Swift, apart from slipping, is that she becomes so focused on the sheep she doesn't notice the drop below.

We arrived at the ledge and started to make a slow approach to the ewe, who was now looking to move along to an even more precarious spot. The sheep's ears were pricked, and she looked around anxiously for an escape down the narrow ledge. Trying to avoid any sudden movements, I gave Swift the quietest whistle I could. She crept three or four yards down on to the lowest ledge and lay there on the steeply sloping rock face, with her back to the sea, blocking the ewe's path. One of Swift's other great assets is that she can make these kinds of movement gently and delicately, without panicking the sheep. The ewe watched her movements intently but barely moved.

While the ewe was distracted by Swift I took the opportunity to inch my way behind her on the blind side. The sheep is equipped with eyes protruding slightly on either side of the head, giving it the ability to have near 360-degree vision. However, it has a brain that appears to be able to interpret the sight from only one direction at a time, usually the direction from which the most threat is apparent.

I was four feet away when she spotted me again. Once more she looked over the edge, as if to work out precisely how she was going to run down a vertical cliff. I sat down, partly to try to appear less threatening, partly because I felt a little unstable. As I regained my balance, a slight movement from Swift again attracted the ewe's attention, and allowed me to sneak another few inches closer.

Having a crook in a situation like this is vital. It's about four feet in length and gives you a reach that the sheep doesn't anticipate. Just waving to catch the sheep's eye is usually enough to turn it in the direction you want. Using the crook, I turned the sheep back up the ledge away from me. I was now within reach so gave Swift the instruction to get to her feet. As the ewe turned her attention once more towards the dog, I made my move. Using the crooked end of the stick, I grabbed her half-way up one of her back legs. I gave a quick jolt back towards me, breaking her balance for a moment and allowing me to grab her chin with my left hand. Dropping the crook, I grabbed the wool on her flank with my free hand. She gave a slight lurch forward, but my grip was good. I had her restrained.

The next bit was likely to be just as tricky. My back was now facing the sea and, although the crash of the sea swell was nowhere near as severe as on the north side of the Point, it still served to remind me that I didn't really fancy dropping in. This ewe probably weighed about sixty kilograms, around seven stone, so lifting her to chest height over the overhanging ledge, while standing on steeply sloping rock myself, was going to take some doing.

Satisfied I'd got myself in the right position, I 'hotched' the ewe up closer to the overhang. With one arm under her chest and the other over her back, I gave her enough of a lift to get her front legs on to the next grassy ledge. After a quick change of holds, I pushed her back-end upwards, hoping that she would have enough sense to try to assist her ascent. She scrambled up a yard, Swift and I followed quickly after her, and secured her again before she had a chance to turn. The next two ledges presented slightly less of a challenge and soon sheep, dog and shepherd were on level ground, two of us at least feeling rather happier for it. The ewe was completely unconcerned – she wandered off up the hill, picking hungrily at shoots of grass and cliff foliage on her way.

The drama seemingly over, most of the walkers began moving away. 'Glad to have provided you with a morning's entertainment,' I thought to myself, rather unkindly. It wasn't their fault the ewe had been so stupid. Out of the corner of my eye, I saw Ernie straining one more time at his lead, letting out a bark-cum-yelp of sheer frustration. I walked back over to him, where Tom, the young lad, was making Ernie a lifelong friend. The drama seemed to have made his day, and he spent a few minutes asking questions. Why was Ernie so keen? Why did the sheep want to jump? And why did it go down there in the first place? That was a question I really couldn't answer. I sensed farming was in his blood and wasn't surprised when he told me his grandfather had run a sheep farm until a few years ago.

With the drama over, Greg rejoined us from the cover of some long grass, where he had been watching the events unfold. This sort of close work has never been his strong point. I walked up the hill towards the Land Rover, feeling quietly pleased with myself. The morning could easily have been a disaster.

The Land Rover bounced violently over a mixture of old rabbit burrows and ant hills as I made my way back up the steep slope. Tom was now a little further on with his parents, but turned to wave me on my way. Standing there with a grin on his face, he reminded me rather of myself at that age.

Unlike Tom, though, there was no agricultural background in my family in Kent. I often wonder what I'd have done for a living if I hadn't spent many of my childhood holidays on what is now the next-door farm to us on the North Devon coast, Damage Barton. Lambing then was heavily dependent on a willing supply of helpers, and as a child I could think of nothing better than spending holidays driving ewes and their newborn lambs from the lambing field back to the barn, or helping with the feeding for long hours in the sheds.

I found something magical about the sights, smells and sounds I encountered: the contented noises of a ewe with her newborn lambs in the sanctuary of the lambing shed on a sodden and windy spring morning, the cold wind fresh from the sea on my face, the burbling stream running through the farm's courtyard. At the age of four or five, they fired an ambition within me that was never to wane.

By my early teens I was spending my days in Devon rolling wool for the shearers, and carting hay and straw bales in from the fields at harvest time. No one was surprised that within a month of my seventeenth birthday, and precisely two days after passing my driving test, I had left home to start work as a Youth Training Scheme apprentice on a 500-acre mixed arable and sheep farm in East Kent.

Life as a YTS apprentice was a little different from the idyllic experiences of my West Country holidays, I soon discovered. I spent the first eight weeks of my apprenticeship standing on the footplate of a corn drill being towed by a tractor driven by my boss, Roland. The corn drill held about half a tonne of seed and fertiliser. My job was to notify Roland whenever it needed refilling.

After two months quietly freezing as I reported the state of play of the hopper in front of me, we were working on the last field when I noticed we were running low on seed.

'Roland, we need to fill up,' I called towards the open rear window of the tractor. Receiving no response I tried again. 'Roland, we're running out of seed.' Still no response.

As we turned again at the end of the field, I climbed up closer still to the cabin. 'Roland, we need to . . .'

At that moment Roland shot around in his seat and shouted: 'David, I know far better than you do how much seed is in that bloody drill.'

Somehow the role I'd performed for the past eight weeks seemed a little less important. Fortunately the farm's shepherdess, Judith, was more willing to make use of my youthful enthusiasm. I'd first met Judith on my initial interview at the farm, a month beforehand. 'Interview' may be a little bit of an overstatement, as on my arrival no one seemed to be expecting me or even know who I was. Roland was busy elsewhere and never appeared. Having parked in the yard and searched in vain for some sign of life, I was ready to give up when Judith had emerged from an impenetrable cloud of dust inside the grain store door. Mask over her face, her long blonde hair laden with dust, she had done her best to make conversation for ten minutes, before disappearing into the cloud once more, lost in the din of screeching conveyors and corn-drying fans.

By the late autumn her grain store duties were over and she was free to pick up on her real passion, looking after the 480 ewes that were scattered in isolated flocks for miles around the farm. Our working relationship was soon blossoming, helped I imagine to some extent by my admiration for her two sheepdogs, Gail and Wee.

Having spent my first couple of months on the farm as no more than a spectator, working with Judith was a joy. She had a refreshingly different attitude, letting me attempt the jobs that I'd been too young to try during those holidays spent in North Devon. During the daily rounds, she even allowed me to drive the Land Rover, sitting there stoically, with both hands braced on the dashboard, white knuckles in view as if anticipating an inevitable collision.

No incident was more memorable than the one that occurred on our first morning together. A ewe had died in the 'hospital' pen at the farm. Determining the cause of death on the farm can be something of a guessing game. In this case, however, the sheep had two puncture wounds on its chest, surrounded by a large bruised and swollen area the size of a frying pan. It was pretty safe to assume this was a bite from an adder.

Whatever the cause of the sheep's death, its corpse had produced the most appalling stench I'd ever encountered. To make matters much worse, as we were loading the ewe into the back of the Land Rover for disposal, the fluids of the dead sheep's stomach suddenly gushed out all over my leg. Despite my best efforts to remove it, the smell preceded me for weeks to come.

Judith had a refreshingly earthy attitude when it came to her dogs too. Once we stopped by the side of a lane to check on a small flock of sheep. On completing the task, Judith's dogs ran ahead to the Land Rover, where they were met by a large collie from the village. Wee, the eldest of Judith's dogs, was in season, and the large collie had nothing but amorous intentions. By the time we arrived back at the gate it was already too late to deter the village Casanova. Judith, completely unconcerned, reached for her flask of tea, 'A basic right of any dog,' she said.

As I settled into my apprenticeship I spent numerous hours in Judith's kitchen, chatting to her and her husband Colin. Their hospitality and company were the perfect antidote to the shock of living away from home for the first time. I'd been on the farm a year when Judith announced that she would soon be taking maternity leave. Immediately my hopes were raised that I might be in line to succeed her. With only a year's experience assisting under my belt, I realised I was being rather optimistic. However, what I lacked in experience I made up for in enthusiasm. And with Judith still available to oversee, I was offered her job – much to everyone's amazement.

The following September Judith's baby arrived in true Judith style, two months early, brought on as a result of her spending the day throwing the sheep into the sheep dip!

As I took over her duties, I acquired my first sheepdog, Kim, a puppy from a litter produced by one of Judith's dogs. Over the next few years Kim proved a fantastic dog, perhaps one of the best I've ever had. I don't know what I would have done during those early days if it hadn't been for the combined wisdom of Kim and Judith.

Judith also kept ten sheep of her own, in the paddock of her parents' house. With her time taken with motherhood, I was more than happy to keep an eye on them for her. My willingness to tend to Judith's flock might have had something to do with the fact that her younger sister Debbie would come out to assist me. She'd generally appear with her face lost down the tubular hood of a snorkel-shaped coat, a garment that had never been fashionable, and never would be. Together we gave the sheep their vaccinations, pierced ovine boils and trimmed hooves.

The romance of it all proved irresistible to us both. Four years later, with Debbie wearing a rather more elegant and glamorous outfit than the snorkel coat, we were married.

CHAPTER FIVE

Trials and Tribulations

The dire consequences of global warming, it appeared, were upon us. On the television and radio the weather reports every night were of flooded rivers, submerged and cut-off communities. In one village only twenty miles away, the pub was half under water and the raging torrent of the river had come up to the top of the ancient stone arches of its bridge.

At Borough Farm the rains were no less relentless. The ground was a saturated sponge, with new springs forcing their way to the surface, and small streams being created alongside the stone banks where they ran down the side of the hill. Even the birdlife seemed fed up with the unremitting greyness of the world. Most mornings I'd pass one of Borough Valley's buzzards sitting on a gatepost at the top of the lane, her feathers sodden. During the summer I'd watched this particular bird, distinctive because of her unusually white plumage, soaring effortlessly on the warm thermals. Now, reluctant to even attempt taking to the air, this proud hunter had been reduced to picking over worm casts.

The weather made the endless routine of winter feeding thoroughly miserable, and nowhere was it more dispiriting than in the fields close to the farm where I had moved a group of 150 lambs. With no hope of them finishing on the diminishing grass, I'd brought them here to give them the additional quality feed they clearly needed each day.

The exercise had turned into a real chore. Each day I had to move their feeding troughs a few yards to avoid the ground becoming poached into a sea of mud. But as the weather worsened, there seemed nothing I could do to stop the lambs becoming caked in it, the hair on their legs and their woolly bellies turning into a mass of muddy clats. Their condition was standing still, at best. As I ran along the line of lambs each day, quickly feeling their shoulders, backs and loins to assess their condition, I became more convinced that what nutrition they were receiving from the trough was being burnt up by the need to produce enough energy to keep warm.

Today, as I stood and watched them, the rain hammering once more on the back of my hood, the whole demeanour of the flock seemed utterly miserable. By now they had taken such a prolonged drenching that their wool was parted lengthways along their backs. The rain was running straight through on to their skin rather than being shed, as it should be in less extreme weather. As usual they ate from the troughs with gusto, but as soon as the food was gone they trudged off to take shelter once more behind a stone wall, or in the hollow inside of a gorse bush.

Even Greg looked miserable. As we left the lambs to the elements, he lifted his feet up in a deliberate manner, picking his way back to the quad bike, his tail lifted slightly as if in an attempt to keep it out of the mud. At least I could console myself with the fact that the other lambs 'at keep' at West Down were faring rather better. Since moving to the fresh pastures they had begun to look as if they were improving. Their wool was bright and healthy, and their bodies had a rounded appearance.

Paul, the farmer whose fields I was renting, had let me have the use of a cow yard. Today, as I penned the lambs up and felt each one over its back, I was relieved to find that two-thirds of them were finished and ready for market – well, almost ready. For the last three or four years sheep farmers have been compelled to belly-shear lambs before they are sold in wintertime. It's a time-consuming and thankless job that involves taking off the wool from the belly of the lamb all the way to the crutch. I understand the hygiene issues behind it, but it doesn't make the job any easier. It is miserable work.

I'd set up my shearing equipment next to the pen in the cow yard. It wasn't perhaps the best choice of location, exposed as it was to the weather from the coast. It seemed to catch the cutting wind with more ferocity than anywhere else on the farm. With the wind biting at my back, I set about turning up each individual lamb, then, bending over double, running the electric shears through the wool on their mud-caked bellies. As if their bellies weren't bad enough, their feet and legs were soiled with cow slurry from the yard, copious amounts of which were soon covering my arms, hands and clothing. Every now and again a lamb would kick out, splattering my face.

At a rate of forty sheep an hour – at best – the 128 lambs in front of me were going to take me another three hours and more. Summer had never seemed so far away.

By the third week of November, tupping was over: the ewes were all served and the exhausted rams had been freed from their annual duties once more. There is never an ideal time for a short break on a farm, but this was as close as it came. So the phone call I got from a shepherding friend in Cumbria was well timed.

I'd got to know Derek when I bought Fern from him a year earlier. He was a Scot, and a well-respected figure in the sheepdog world, both as a trials handler and a breeder of good strains of sheepdog. A few months after Fern had arrived in North Devon, I'd travelled up to his farm to mate Swift with his outstanding sheepdog Sweep. The result had been the charismatic – if problematic – Ernie.

Derek had called to let me know he was running a 'hill trial' on the fells above his farm. 'It'll be a real test for you, if you can make it,' he said. The fell course over which Derek ran his trial was infamous, but for that reason attracted some of the country's best sheepdog handlers. I was flattered to be asked, but foresaw problems persuading the family to let me have a weekend away on my own.

A day out at Exmoor Zoo proved the perfect bribe for the children. Debbie, I feared, would prove a different matter. Derek's invitation extended to Debbie too, but she didn't like to stray too far from the children when they were still so young, and was generally underwhelmed by the prospect of two days standing on a windswept Cumbrian hillside watching sheepdogs. 'You go, the break will do you good,' she said, understanding as ever.

My only decision now was which dog to take – Greg or Swift. Swift eventually got the nod because of her superior hearing over the sort of distances we'd be working at in Cumbria.

It had been a good friend and fellow local sheep farmer who had got me interested in sheepdog trials. Andrew farmed the National Trust land a few miles along the coast. A man of few words, he'd made an unforgettable impression the first time he came to visit us when we moved to Borough Farm, seven years earlier. He and his wife had come down to the farm to introduce themselves and stopped for a cup of tea. Helen had done most of the talking. Andrew had sat back in an easy chair, saying very little and looking just a little glum.

His expression changed only when Debbie brought out the cake tin. For a moment his face brightened, only to break into a frown again when he was offered a slice of coffee cake. 'Haven't you got any chocolate?' he grumbled.

We had become firm friends nevertheless, in part because of our mutual interest in the dogs with which we worked. Andrew had been competing in sheepdog trials for some years and introduced me to the sport. A year or so after we met, the National Trust organised a sheepdog trial for tenant farmers at Ysbyty Ifan near Bala in North Wales. Without my knowledge he'd entered us both to compete.

'Thought we should represent North Devon,' he announced, concise as ever, but with the trace of a mischievous smile on his face. Springing such surprises would become a habit in the coming years.

We'd headed up there with two dogs each, in my case Greg and my previous dog Rush. The downside was that Greg was only fourteen months old, barely more than a pup, while Rush was singularly unspectacular in most aspects of her work and spent most of her time lying down just watching the sheep.

It was entirely fitting that the event was taking place in this corner of North Wales. It was there in Bala, almost a hundred years ago, that the very first sheepdog trials were run. The trial the nineteenth-century shepherds faced was designed to replicate the day-to-day challenges of a working sheepdog's life in a competition environment. Ever since, most courses have conformed to roughly the same format.

The object is to send your dog to collect five sheep that have been released at a distance, usually more than 250 yards away. With the handler still standing at his post, the dog must first start to move or 'lift' the sheep in a calm and controlled manner. The dog must then push or 'fetch' the sheep back to the handler, before driving them around a triangular course. This 'drive' should take the sheep through two sets of gates and back into a 'shedding ring' marked on the grass in sawdust. There two of the sheep are separated from the rest and held apart to the judges' satisfaction. Finally the quintet of sheep are reunited and – usually much against the sheep's better judgement – persuaded into a small pen. Each of the six sections of the run is allocated a number of points, totalling one hundred. If a dog makes a mistake during any of these sections, some or possibly even all of these points are deducted. In competition the winner is the handler who has got around the course in the allowed time and has the highest remaining points score.

This was the task that had faced me as I came to stand at the starting post at Ysbyty Ifan. I'd felt under-prepared and utterly intimidated. I'd rather hoped that the hundred or so people in the audience suddenly felt the urge to go to the beer tent. To my huge relief, however, when their individual turns came, Greg and Rush both managed to retrieve the sheep and drive them, albeit haphazardly, around the course. They even managed to get the sheep to pass through the odd set of gates on the way, although not necessarily in the right direction. I'd avoided total humiliation and that was all that I could have asked for at that point.

Yet something happened that day to whet my appetite. In part it was great to see the control and ability that some of the best handlers demonstrated. But I also felt a challenge being set for me. This was my specialist subject and I should be able to do as good a job at it as anyone else. Whatever the reasons for it, the seeds had been sown and I'd been trialling off and on ever since.

Derek's event was, however, likely to be a completely different experience.

In North Devon the taste of late autumn was still in the air, but three hundred miles to the north winter had well and truly arrived. The burns were in full flow and the winds cutting across the fells chilled you to the bone.

I spent the evening at Derek's kitchen table, talking sheepdogs and growing ever more alarmed by the list of handlers who were dropping by to confirm their entry for the following day. The field assembling there was intimidating. I recognised the names of the other competitors from the trials results regularly published in the *Farmer's Guardian*, the livestock farmer's paper of record. There were national and international champions. It didn't help me get a good night's sleep.

At eight o'clock the next morning I headed to the course, set on the steep hillside at the foot of Lonscale Fell. The sun was late in lifting above the fell that loomed high above us. In the gloom, huddles of unfamiliar faces quietly discussed the pitfalls of the course slowly revealing itself. The cutting wind blowing from the north made the eyes stream, but no one complained or even mentioned the chill. Most of those gathered here spent a good deal of their working lives in such conditions. Today was recreation and it was going to take more than an icy blast to spoil it.

I'd stood on this hillside once before, when I'd taken Swift to mate with Derek's dog Sweep. It hadn't been an experience I was likely to forget easily. I'd set Swift off to gather some distant sheep. It was after she'd run a couple of hundred yards in completely the wrong direction that I realised that, for the first time in my shepherding life, I'd forgotten my whistle. As I frantically searched through my pockets, Swift had jumped a fence and set about rounding up the neighbour's flock. Unable to find a spare whistle, to my acute embarrassment I'd had to explain to Derek what her stop command sounded like. Thankfully, she'd responded to his whistle and eventually returned. Derek clearly enjoyed reminding me of the incident and did so at regular intervals. As the competition got under way this morning, I tried my best to put the memory to the furthest recess of my mind.

Today's course had been fiendishly laid out. The right-hand route was very difficult, with the dog first having to scramble 150 feet down a steep bank, over a stream and up on to the hillside, before crossing two gullies that ran up the side of the fell. However, the left-hand option was, as far as I was concerned, nearly impossible.

It was soon clear that no one was finding the going easy. One highly experienced handler lost sight of his dog as it crossed the stream in the valley below. When his dog eventually reappeared five minutes later he did so minus any sheep and from the opposite direction he should have, surprising his handler by suddenly appearing behind him.

From the point where the spectators had gathered, half-way along the course, it was hard to see the test in its true perspective. When it came to my turn to go to the starting post I saw the full, daunting extent of the course for the first time. The ravine in front of Swift was so deep it disappeared out of sight. On the other side of the fell, what had looked to be a pair of insignificant dips turned out to be gullies deep enough to lose not just a dog but an entire flock of sheep inside them.

Three-quarters of a mile away I could make out the five sheep being let out of the pen. At that distance there was no way Swift could see them. I sent her off to the right-hand side of the course. She disappeared down the steep bank but thankfully reappeared, crossing the stream at the bottom. A couple of commands from me and she was heading in the right direction.

Seeing Swift well on her way, I looked up the fell and realised the sheep had disappeared over the horizon. Swift was getting close to where they had been – hopefully, close enough to see where they had gone to. Soon she too was an ever diminishing black speck on the horizon. Now it was all down to her.

I gave several loud blasts of encouraging whistles. On this hill trial the hardest part of the run was getting the dog to physically retrieve the sheep. As the seconds passed, my mind raced with dark thoughts. A triallist's greatest fear is that his dog won't 'lift' the sheep and he'll be forced to go and collect it from the course. The humiliation here would have been even greater because it was a fifteen-minute walk to the point where the dog was fetching the sheep. 'Surely I haven't come all this way to be shown up like this,' I said to myself.

Then at last, and to my huge relief, there was some movement on the horizon. Squinting, I could just about make out the form of sheep against the dull browns of the autumn fellside. At first I couldn't see Swift, but the movement of the sheep meant that she must be there. When she appeared, I gave her a loud left-hand whistle command, telling her to steer the sheep towards the first obstacle, a large rocky outcrop, further up the fellside. The distance was such that it took a full second for the sound of the whistle to reach her. But she responded well, and between us we kept the sheep under control and roughly on course.

Ten minutes later, with only a few hiccups along the way, I shut the gate on the five sheep to finish the trial. I called Swift over to me and gratefully rubbed her neck, praising her quietly.

Simply getting around a course like that gives you a sense of achievement. But on my way back I walked past John Thomas, one of the country's most successful sheepdog handlers for more than thirty years. Handlers aren't men of many words but his brief 'Well done' gave me real satisfaction.

For an hour or two I was actually leading the competition. Predictably, in the afternoon I was overtaken by some really good runs. But I wasn't at all disappointed. I felt I'd proved something to myself by getting Swift around the course. However, it wasn't long before I was brought back down to earth.

I called Debbie that evening feeling fairly pleased with myself. It turned out she'd been having a less enjoyable time than me. Whenever I was away for a day or two Debbie would give the dogs as much work and exercise as she could. But while the other three dogs had been behaving impeccably, Ernie had been up to his familiar tricks. While out on a run with her, he had slipped his collar and begun rounding up a few sheep in the field below the farm. Nothing Debbie said or did would distract him from the job. It had been nearly half an hour before he'd worn himself out sufficiently for her to catch him.

Debbie understood my fascination with sheepdogs, and the important role they played on the farm. Generally she enjoyed having them around, but at times like these she felt less well disposed to them.

'Why can't we have a nice normal dog we can take for walks like everyone else?' she said, more than a hint of frustration in her voice.

I'd clearly been spending too much time among sheepdog handlers. Listening to her, I couldn't help thinking it sounded like promising work on Ernie's part. 'He didn't let any go, did he?' I asked, realising the insensitivity of what I was saying almost as I said it. The silence was deafening, but eventually came to an end with an exasperated sigh from Debbie.

'Next time you go away you can take that stupid dog with you.'

Winter

CHAPTER SIX

A Sheep's Worst Enemy...

The morning sky was a flawless milky blue but the air was so cold it froze my breath the moment I stepped out into the yard and headed for the dogs' run. The drier, chillier weather had come in the previous day on a front of high pressure that was spreading in from the east. The dogs didn't need much pampering in winter but, with the temperature dropping fast the night before, I'd added some extra straw bedding and rearranged the kennels so that they faced away from the biting winds. It turned out to have been a good decision. Overnight the water bowls in the kennels had frozen solid, and I had to break through a couple of inches of ice to draw them a bucket of water from the butt in the yard.

The older generation of farmers still debate over whether North Devon's worst winter was in 1946 or 1963. If conditions were worse in 1946 than they were seventeen years later, the area must have resembled Siberia. Some farming friends of mine were snowed in without electricity for eleven weeks on their Exmoor hill farm in 1963. It seems incredible now, but as a direct result of that they sold up and moved to South Devon in search of a warmer climate. Whichever of the two great winters was the colder, the indisputable truth is that forty years on a 'healthy' freeze-up is a rarity.

The arrival of some colder, sharper weather was a relief in more ways than one. The relentless rain of the past month or so had become utterly depressing. More importantly, however, the warmer, wetter weather of recent years seemed to have allowed bacteria and disease to survive – and even thrive – on the farm. A good cold spell might suppress some of the bugs that were about.

I'd now resorted to housing the 150 mud-caked lambs I'd been feeding in the fields. I'm sure they'd felt as relieved as me when I brought their prolonged soaking to an end. With these lambs now well bedded on straw and eating an intensive diet of concentrate feed, they'd hopefully be 'finished' in time for the New Year. For the last few mornings, however, I had been treating a few of them in the shed for lameness. It is sadly a common problem with sheep, usually caused by a variety of bacteria and exacerbated when sheep are living in close proximity to each other. There's an old adage that the sheep's worst enemy is another sheep and it's very true, particularly where foot diseases are concerned. Put a group of sheep together in a confined space and problems can easily spread from one to the other.

I'd started applying the standard treatment, which is to trim the hoof radically, wipe it clean and spray it with a commonly used antibiotic spray. Usually this works. Within a few days of starting the treatment, the hoof is hardened over, well on the way to recovery.

The dogs ambled about the yard and sheds, while I spent half an hour in the comparative warmth of the buildings, filling up the feeders and strawing the pen. But as I cast my eye over the lambs I soon realised that not only were there a good many more lame sheep this morning, but those I'd marked as having been treated were no better either.

A little concerned, I turned over one of the previously affected lambs and re-examined its feet. Small, red, wart-like blisters were now obvious at the top of the hoof. I used the foot trimming shears to cut back the hoof, this time from the top. This now revealed an infection, spreading between the live section of the foot and the horn, parting the horn away. Perhaps I had missed it during the previous treatment or perhaps the symptoms had now progressed; either way the problem was obviously not the common footrot that I had been treating but the more virulent contagious ovine digital dermatitis, or CODD.

I had a good idea how it had got on to Borough Farm. About five years ago while the lambs had been wintering on a dairy farm, the bovine equivalent of the disease had been identified in the herd of cows there. Some of those lambs had been a bit lame when they had returned home. Although I wasn't aware of it then, I was by now quite convinced that's when the problem arrived on the pasture here. This was precisely why I felt we needed a cold winter. In the absence of anything like a freeze-up in recent years, the bug had persisted, periodically resurfacing, particularly at weaning time when the lambs are stressed and prone to problems. It was probably now on the farm to stay. This, however, was the first time I had encountered CODD in the confines of a shed, where it had the potential to be far more easily transmitted. It was a worrying development, to say the least.

Every shepherd needs a basic veterinary knowledge, although he can't possibly know everything – *The Veterinary Book of Sheep Farming* runs to 680 pages. Fortunately many of the ailments listed within this door-stopping tome are rare, or regional problems, some unheard-of in North Devon. Unfortunately, the problem with CODD was that it was a new threat. It had been identified as being present in sheep only within the previous five years.

By word of mouth, I'd heard of an effective cure. The farmer who had originally been affected by the problem told me that he had found only one antibiotic of any use in the treatment of his cows. The powder was prescribed to him for use in a footbath through which the cattle would walk. He, however, had discovered that by mixing it in a small hand sprayer he could apply it directly to the affected foot – at a fraction of the cost and with even more effective results, it seemed.

This was something I had to try and nip in the bud, so I went back to the house to call the vet. He said he'd leave a tub of antibiotic powder ready for immediate collection. I shot up to the surgery then headed home ready to start applying it straightaway.

I penned the flock tightly and started turning up the sheep individually, scraping away the accumulated dung with the shears, then inspecting each foot. Some were as yet completely unaffected, but a good many were showing some sign of the disease. I pared each one, removing all the loose horn under which the infection traps and festers. There is a thin line between trimming too much and too little. By not trimming enough you can leave cavities under which the bacteria can continue to multiply. By trimming too much the foot can bleed. Although I tried hard to avoid over-trimming, inevitably I drew some blood as I worked to remove the rancid diseased tissue from the live hoof underneath.

Each time I'd completed this part of the operation I plucked a little belly wool and used it to wipe clean the infected area. I then applied the antibiotic from the hand sprayer.

It was slow, unpleasant work. After nearly three hours I stood up and straightened my back. The wind was still cold, but I had certainly warmed with the work. The infected lambs looked no better for the treatment; many of them were sore-footed where the infection was exposed. It would be several days before I could hope to see an improvement.

As I broke the ice on the water tank and washed the worst of the stinking matter from my hands, a robin flitted on to a hurdle near me and half-heartedly delivered a burst of shrill song. I didn't feel in great tune myself.

As a shepherd I detest the sight of lame sheep. Apart from the obvious pain and suffering it causes, it reflects badly on myself. I take it as an affront to my abilities as a flockmaster. Lameness is, however, an inevitable consequence of keeping sheep in close proximity to one another, in a warm and wet climate. Shepherds across the country must spend thousands and thousands of hours treating the problem. The three hours I'd spent this morning would be only a fraction of the time I would spend on it during the year, although I hoped some of that time would be passed on a slightly warmer morning than this.

The healthy cold spell proved fleeting. Within a few days the wind swung viciously back in from the Atlantic, bringing with it a return to the relentless wind and wet weather of November. With the winter routine of feeding in the fields as unremitting as the rains, it didn't do much to improve my spirits.

At the beginning of December I had moved the bulk of the flock to the steeply sloping banks on the opposite side of the valley to the farm. These fields have only a thin layer of topsoil. This allows the water to drain freely through the shillet underneath and in turn allows me to keep the sheep as dry as is possible in these conditions. The fields' other advantage is their proximity to the woods at the lower field boundary, where the flock can get excellent shelter from the wind and rain.

The grass was now dormant for the winter, so the ewes were completely reliant on the big bales of silage I was now taking to them each day by tractor. During the summer I'd made four hundred bales of silage, enough to see the flock comfortably through the winter.

A new routine had taken shape. Having seen to the lambs in the shed each morning, the second task of the day was to take three or four bales out to the various feeders that I'd positioned in each field in a position easily accessed from the road.

As I swung open the door to the shed, Swift was already sitting in the tractor's seat waiting to go. She never required any encouragement to join me and only needed to see me approaching the tractor to jump into position. Greg wasn't quite so enthusiastic. He hated mud and, being a much larger-framed dog, found the jump into the tractor rather more difficult. As usual this morning I left him behind. 'Move over, Swift,' I said as I climbed in and turned the ignition.

After I loaded a bale of silage on to both the back and front of the tractor, we made the short trip around the roads. The small knoll where the feeder stands catches the wind like nowhere else on the farm. With the tractor I lifted a bale over the feeder, and forced the door open against the wind. With my hood rattling around my ears like a tarpaulin in a gale, I cut back the black wrapper. As I pulled the covering off the bale a shower of grass stems and seeds drove into my eyes. I blinked heavily and wiped them with the sodden sleeve of my coat, but all I succeeded in doing was leaving mud across my cheeks.

Swift joined me from the tractor. Even she seemed reluctant to get out into the storm, but on these occasions I needed her more than ever. Fresh silage has a distinctive, almost sweet smell, which draws the ravenous ewes to it. However, they have an unfortunate habit of putting their heads into the feeder before the bale has been dropped by the tractor. Sure enough, a few ewes that had been sheltering close by spotted the arrival of the fresh bale and made a beeline for the feeder, thrusting their heads underneath.

I can't say that all sheep are stupid, I haven't met them all. But the ones that I have encountered tend to be on the lower side of dim. This bunch were completely oblivious to the fact that a quarter of a ton of bale falling from that height would drive them into the bars of the feeder with such force it would undoubtedly kill them.

It was a routine for which Swift needed no commanding – she dived in between sheep and feeder, sending the suicidal ewes jumping back to regroup a few yards away. They stood indignant, contemplating another attempt on the feeder. Swift stood facing them motionless, her hard eyes staring intently at the small group. Her message was clear. The sheep stayed put until I could drop the bale in place.

The relentless bad weather was taking its toll on the ewes. That morning as Swift had brought up the rest of the flock from their shelter by the woods, an older ewe had hung back from the rest, then turned to face the dog. It was a typical stance for a sheep that was struggling. It was not that the ewe needed any particular medical attention, it was more her age and the harsh weather. She was simply too weak even to climb up to the feeders at the top of the incline. I resolved to make an extra trip to take her back to the farm later in the day. Just to be on the safe side, I'd also give her an extra worm treatment in case it was liver fluke. The likelihood was, however, that warmth, some extra nourishment and a dry bed were all that she would need to see her through to lambing.

It wasn't an option available to the ewe I came across a short time later. I'd been drawn to the dead bracken on the steeply sloping bank in the corner of the field by the sight of two buzzards emerging from the undergrowth. The object of their interest was soon apparent. A ewe's carcass was half-eaten and lay where she'd succumbed to the foul weather overnight. The law says it's my duty to dispose of her but I was in no hurry as I trudged up the hill. My loss represented a rare winter bounty for the scavengers of the Valley.

CHAPTER SEVEN

December Days

It was now a week since I'd detected the CODD in the fattening lambs and it was obvious that, far from lifting, the problem was getting worse. I had moved a footbath into the pen, and was running the lambs through each morning. I hoped this would help to harden their feet, and at least stop the disease from spreading. Although some of the originally infected lambs did now seem to be improving, there were still a few more cases each day despite my efforts. This had ceased to be a problem of normal proportions – it was becoming an outbreak that could potentially affect all the lambs in the pen.

The lambs wouldn't die of it, but neither would they thrive. It hurt my pride as a shepherd that I had let this problem creep up on me, and the thought of the lambs at the back of the shed nagged at me all day. I spoke briefly to the vet again over the phone. He agreed that the treatment I was giving was correct. There was an option to give an antibiotic injection but, like me, he wasn't keen on the idea. Quite apart from the expense, farmers and vets alike are trying hard to avoid introducing drugs like this into stock that is eventually destined for the market.

He suggested increasing the bedding straw, and separating off the affected lambs. In retrospect, I should have taken out the lame lambs in the first place, so as not to infect those around them. But I'd wrongly assumed that a couple of days' treatment would cure the problem. The vet's closing remark didn't do much to improve my mood. 'I've seen outbreaks where the worst cases had to be slaughtered, but I'm sure it won't come to that,' he said rather gravely.

I carried on with the same routine: a daily run through the footbath, turning up and treating affected feet, and bedding the lambs on copious amounts of straw. As I settled into the miserable routine the same thought kept rattling around in my head. Not only were the lambs suffering, but also they couldn't be sold. I had to get on top of this problem.

With work mounting up on the farm, I hadn't been over-enthused when I got a phone call from Chris, the organiser of the shearing gang with whom I worked. Each May and June I teamed up with Chris, and another farming neighbour, Geoff, for the shearing season. Our round consisted of a list of customers that had been passed on from shearer to shearer over the last thirty or more years.

Chris was the longest-serving member of the gang. He'd been contract shearing for nearly twenty years. Geoff had joined Chris ten years ago. I was the newcomer, having been involved only sporadically for the last five seasons.

Some of the customers also liked to have us back during the winter, to dock their ewes. Chris had been asked to dock six hundred or so for a local farmer, Michael. 'If you could help us tackle them, he wants them done in a day,' he said. I couldn't really say no, particularly as I knew Geoff was likely to be already working eighty hours a week, milking cows, and Chris, who farms on the other side of West Down, had no more spare time than I did. I'd agreed somewhat reluctantly.

On a dry but dull mid-December morning I set off for the first venue, a field somewhere near Berry Down. I say somewhere, because Chris was not always precise with his directions. Michael kept his flock in groups of one or two hundred, spread out over a ten-mile radius from his farm. Chris's directions were basically that I turn right as I passed through the village, then make a series of left and right turns, down some typically narrow Devon lanes. The first flock, he said, were in a field on the right. 'You can't miss them,' he reassured me.

It wasn't long before things went awry. After five years of sterling service, my fifteen-year-old, B-registration Land Rover was on its last legs. If it had been a dog it would almost certainly have been put down. The gearbox was in such a bad way that third gear no longer existed. The passenger side door had a nasty habit of flying open whenever I took a right-hand turn at speed. After a shaky start and a mile or so of coughing and spluttering from the engine, it had just about settled into a rhythm. No sooner had that worry lifted than I was aware of another problem – Chris's directions didn't seem to work.

I'd followed his instructions to the point where I should turn right, but had found no sign of my fellow shearers on either occasion. So now I turned the Land Rover and its small trailer, containing the shearing equipment, down an unmade track. As I headed up steeply through some woods, it became obvious that the track was an access road to a farm, and before I knew it I had entered the yard. It seemed rude to just turn and drive away, so I decided to try to get some directions.

A weathered-looking farmer, in an old beige stockman's jacket, appeared from a low barn and ambled across the yard to meet me. I recognised him from a few brief meetings at the local markets. His name was Archie. During my previous encounters with him, Archie had seemed a slightly dour character. Today, however, his face seemed to lighten at the arrival of a visitor in his yard.

'How do,' he said.

I guessed that it had been some time since anyone had dropped in, particularly when he started striking up a conversation. Inevitably, the subject was the one that was dominating all farmers' minds at the moment – the wet weather. Archie then set off on a procession of meandering stories about flooded crops and mudded fields. 'Worst rain since Father's time,' he said, shaking his head slowly.

I didn't want to appear rude, but I was anxious to find out where Chris was expecting me. Yet every time I thought I was going to get the chance to ask some directions, Archie found a new aspect of winter farming to muse upon. Each theme was invariably illustrated by a lengthy anecdote. I had no option but to nod repeatedly and make confirming murmurs. I kept the words to a minimum for fear of jogging his memory about any other stories he had saved up.

Eventually, Archie had to draw breath and as he did I seized my chance to explain my presence in his yard. The moment I mentioned I was on my way to dock Michael's sheep, however, he was off again.

'Docking? See what you make of this, then,' he said, ushering me to his workshop to view the shearing machine he'd purchased at a recent farm sale, for 'not much money at all'.

It was half an hour before I managed, without causing offence, to make an exit from Archie's yard. Now I was running really late. Archie thought he knew where Chris meant me to meet him and, to save me returning back through the village, directed me up through his farm tracks, from where I could pick up the right road. Unfortunately he too had forgotten to be overly specific with his instructions. The track I was supposed to take, I later realised, was the second on the right heading out of his farm. Instead, I took the first.

As I turned around yet another bend the overhanging brambles were scratching violently down either side of the Land Rover. Soon I found myself heading down a steep mud track. The tractor ruts on either side were deep enough to make the underside of the Land Rover scrape heavily on the ground. The trailer, its wheels too small to touch the bottom of the ruts, was being pulled along behind like a toboggan. I had no hope of reversing out now and, as I rounded another corner, to my disbelief I found the track completely flooded – judging by the submerged hedge on either side of it, to the depth of about three feet. I was stuck.

Archie was very helpful when I turned up back in his yard on foot. He came out with a tractor and hauled the Land Rover and trailer out, somewhat amused that I could have been quite so daft.

'I never go down there in the winter,' he said. ''Tis always flooded.'

It was another hour and a half before I eventually found Chris and Geoff, no more than a mile away as the crow flies. Both of them were philosophical about my far from punctual arrival. 'Never mind,' Chris said in his gentle Devonshire tones, waving away my apologies. 'We've done a hundred and twenty, the next five hundred shouldn't take us too long.'

My lateness had become something of a running joke within the shearing gang. I could have told them about Archie and the flooded road, but I knew the protestations would have been pointless. Today had merely confirmed my reputation. As we settled down to work, Chris and Geoff couldn't resist making mileage of it all.

'We'll have to start your funeral without you,' said Geoff, hardly lifting his head.

Borough Valley is home to an impressive diversity of woodland birdlife. During the wet and windy weather of recent weeks, however, almost every winged creature seemed to have been sheltering away, hidden from view. Now, with another break in the bad weather, it seemed they were making up for lost feeding days. In the yard, ten or so chaffinches and three grey wagtails picked hungrily over a handful of spilled lamb feed. On the old hay turner that lay, overgrown with nettles, by the sheep shed door, a dunnock was warbling away.

Debbie and the children had put up a bird-table not far from the kitchen window. An ever increasing throng of great and blue tits had flitted across from their home in the conifers on the western slopes of the Borough Valley. Along with their other 'animal' jobs on the farm, Clare and Laura were in charge of refilling the bird-table's supplies of nuts and seeds. They were being kept busy. With the weight of food being consumed each day, it was a wonder that any of the local birds could still get airborne.

Perhaps the only unwelcome sight was the sparrowhawk that swooped fast and low past the bird-table each day, sending the smaller birds darting for cover. It's a magnificent bird to glimpse in full flight, but perhaps it's becoming a little too common. Sparrowhawks are raptors and live on a diet of chaffinch, sparrow, meadow pipit and the like, eating up to three small birds a day. They have been heavily persecuted in the past, but are now clearly thriving, with the current population reputed to be around 64,000. That equates to the eating of 192,000 small birds a day – which doesn't sound sustainable to me. I don't know what the solution is, but sparrowhawks must be partly responsible for the decline of the small birds nationally.

This Christmas the farm's bird population was going to be expanded still further, however. Clare had decided that the next addition she required to her menagerie was some hens. Once set upon an idea, Clare tends to be rather focused. Every time I heard her call 'Dad' in a slightly cautious inquiring tone, I knew the next line was likely to be, for the umpteenth time: 'When can I have some hens?'

So my December list of jobs had now been expanded to include building a chicken house and run. As if this was not enough, I had to achieve this – and the subsequent smuggling in of half a dozen chickens – without the oldest and most eagle-eyed of my children noticing. It was actually a welcome distraction from the routines and concerns of the farm, the problems with the lambs in particular.

The children were getting more and more excited about Christmas. A few days earlier, with the cottonwool beard firmly glued in place and the final smearing of talcum powder dabbed on to my cheeks, I had jumped out of the Land Rover ready for the annual challenge that was persuading a group of two- and three-year-olds that Father Christmas really did exist.

I'd played Santa at the local Busy Bees playgroup for the past seven years, and I'd begun to think that I was pretty good at it, although I'd come close to being exposed a couple of times. One year a particularly curious three-year-old girl almost pulled off a large chunk of my whiskers. But in general I'd got away with it – even with my own girls. Laura and Clare might have noticed Santa's black leather boots bore more than a few traces of sheep dung that had been covered with a bit of boot polish, but both had gazed up into Father Christmas's eyes without realising it was their father.

This year it had been Nick's turn to be given the treatment. I made my usual entrance in my full, scarlet Father Christmas outfit, calling 'Wait there, Rudolph' through the door, then delivering a few hearty 'ho ho ho's.

As I'd begun working my way through two stacks of presents, some of the children had looked a little apprehensive. Nick too had appeared nervous as he approached. When I'd asked him his name, his whispered 'Nick' had barely been audible. His voice had become noticeably clearer when I asked him what he wanted for Christmas, however. 'A tractor and a trailer,' he said, smiling.

The visit ended with a festive song from the children. I called out for Rudolph to get the sleigh ready, then left feeling pleased at now having fooled all three of my children. A couple of days later, though, I'd heard the news that my act hadn't been quite as deceptive as I'd imagined. 'Why does Father Christmas have a Land Rover?' one bright young lad had asked the play-leader after I'd made my exit.

I'd begun to dread the walk to the sheep shed. It was now two weeks since I'd begun battling the lameness but, as I set about the regular foot-bathing and treatment routine, I was beginning to wonder where it would all end. Days are short enough in December without the extra hour I was spending most days in the shed. The last couple of days had been so busy that I had given the lambs no more than the briefest of glances.

Perhaps it was the hint of wintry sunshine visible through the clouds that put me in a more positive frame of mind as I entered the sheep shed this morning. But as I walked slowly through the pen, I dared myself to believe that there were definite signs of improvement at last. Most of the infected lambs were still walking gingerly. When I turned them up, however, there was no question that the feet were hardening, and the raw red inflammation so obviously causing the pain had now mostly disappeared. I checked thoroughly through the pen, and could find no new patients. For the first time I felt that the end might be in sight.

Christmas morning began with the usual dilemma: to slip out before the children woke up, or wait for them to open their presents before doing a quick whiz around the farm. But before I had a chance to make the decision for myself, three children arrived on Mum and Dad's bed, stockings in hand, tearing at wrapping paper and excitedly showing off what Father Christmas had brought them.

Nick's face was a portrait of pure happiness. His second parcel had contained a small digger. From the moment he opened it, everything else was forgotten. At one point he drove the digger over my head, wrapping my hair in its revolving wheels. It took Debbie to release it with a pair of scissors.

Laura was delighted with her 'Furby', a really irritating furry creature that wriggled its ears as it screeched its way through a dozen or so annoying phrases.

'That's lovely, Laura, perhaps you'd like to take it into your bedroom to play,' I said more in hope than expectation.

Clare flicked through a pony book, obviously expecting something feathery but not quite liking to ask.

'We couldn't get you any hens,' I teased. 'So we got you some eggs instead – they're in the fridge.'

Clare looked a little cross.

'Don't be so cruel,' her mother consoled her. 'Get dressed and go and look in the paddock.'

Clare shot out of the room, and a minute later was thumping downstairs. I'd just drained a quick cup of tea and was heading out of the door as she returned.

'Look, look,' she called excitedly, clutching her first egg as though it was the crown jewels. I left Debbie with a turkey to cook for lunch, and an egg to boil for breakfast and then divide between three children.

The dogs greeted me with no more or less excitement than any other morning. The girls had bought them each an individually wrapped chew, which Clare had placed in their runs on her way back from the chickens. Needless to say, nothing remained of any of them by now. I let them out, and they ran playfully across the yard. Swift veered off towards the tractor and looked hopefully up towards the door. She was out of luck. Last thing on Christmas Eve I'd filled all the silage feeders in preparation, so all that was needed today was just a quick check around the sheep and a bit of strawing in the shed.

I walked in and tossed a few wodges of straw into the pen. The lambs ran to the opposite corner, but one or two made a skittish charge back. It was the first real positive sign they'd given me for ages. They picked over the straw as I spread it liberally around, their feet rustling through the dry clean bedding. Their increasing contentment was mirrored by my own state of mind. The Atlantic winds seemed to have driven away the rainclouds of the past few days. The yard robin emerged from a foray into the lambs' feeder, and began singing enthusiastically from the top of the straw stack. This time I was able to appreciate it.

I left the older dogs milling around the yard, where I could see my parents had arrived laden with presents. I headed for the house, church and the rest of Christmas Day, a mild spring in my step for the first time in a while.

CHAPTER EIGHT

Taking Stock

A phone call from our biggest customer put paid to any hopes I'd had of gently easing myself into the New Year. The good news was that Lloyd Maunder were ready to take a consignment of 120 lambs from those I had 'at keep'. The bad news was they wanted to collect them at six in the morning on January 2nd.

So by lunch time on New Year's Day the Land Rover was struggling its way up the lane, pulling a trailer full of shearing gear and mobile sheep gates, bound for West Down. The day was murky, the sky devoid of any colour that wasn't a shade of slate grey. In the fields at West Down, only the occasional cawing of rooks broke the sleepy stillness. By three o'clock in the afternoon I had sorted the lambs and attached plastic ear tags to them, something I have to do by law before they can leave the farm. The light was already beginning to fade as I fired up the generator to start belly-shearing.

Shearing in an exposed, open field on a cold winter's afternoon is no one's idea of fun. It wasn't long before I'd grown heartily sick of the wind cutting into my back where my shirt had ridden up. I'd only completed forty before the light became too poor to work, and I resorted to rigging up the lamp I'd brought with me. In the semi-darkness, it was hard to stop my mind drifting away from the dreary routine. As a new calendar year got under way, it was inevitable that I found myself taking stock.

It had been seven years now since Debbie and I had moved to Devon. During the early years of our marriage, I'd worked as an employed shepherd in East Kent. We'd bought ourselves a house, and welcomed Clare into the world. It had been my parents who had maintained my childhood links with North Devon. A few years before Debbie and I married they had retired and moved to Mortehoe. Each time I visited the county with Debbie and Clare, I felt inspired by the beauty of the countryside.

When Clare was eighteen months old, the opportunity arose to rent land at Borough Farm. As well as the obvious attraction of living in such a landscape, the idea of farming in a livestock area of the country, rather than the predominantly arable part of Kent in which we then lived, was hugely appealing. We sold our house, and made the move across country. It was a decision which we had never for a moment regretted. The opportunity to run our own business, control our own destiny, and bring up a family in such a wonderful part of the country was a chance in a million, a privilege that few families have.

The first five years had been tough, but slowly, steadily, we'd built a life for ourselves. The final piece in the jigsaw, from a farming point of view, had come when the National Trust gave me the chance to rent 260 acres at Mortehoe, 200 acres on the Morte Point peninsula and another 60 acres at the fields of Town Farm behind the village. Since then my flock had grown from around 500 ewes to nearly 900.

The year just drawing to a close had been similar to the preceding six – unspectacular, but successful enough to keep us going. Last year's lambs had sold pretty well, averaging a pound or two more than the previous year, which in the present climate was about as much as we could hope for. Debbie and I accepted that we weren't going to become millionaires through sheep farming – but we only had to look at the childhoods Nick, Laura and Clare were enjoying to realise that we were building a life for our family that was beyond value.

Since Christmas, Clare and Laura had been spending hours looking after their growing menagerie of animals. They seemed to be making ten trips a day out to the hen hutch, no matter what the weather. They'd run back to the house delighted if they'd found two or three eggs, mortified if there were none. Nick was proving equally difficult to keep in the house, heading out to the yard to cast an excited eye over the tractor at any given opportunity.

All this meant that Debbie had her hands full, of course. Looking after the three children, running the house and helping me when required was job enough, but she was also doing a couple of evenings nursing a week. There were times when I didn't know how she did it. I hoped that, like me, she felt it was worth all the hard work.

It was pitch black by the time I'd finished the last of the lambs. My back was so numb from the cold that straightening up proved a real challenge. Getting out of bed at 4.30 the following morning was going to be a painful process. At least I could console myself with the knowledge that the lambs were now ready to be loaded straight on to the lorry, leaving only 250 on the farm to be sold. Which was just as well – the first of the new season's arrivals were due in a little under eight weeks' time.

Despite the remote location of Paul's fields, the lorry arrived within a minute or two of six the following morning. By half-past, the lambs had been loaded and the lights of the lorry were winding their way through the darkness, up the lane on the other side of the valley. Inside the Land Rover, I waited for the first chink of daylight then checked on the remaining lambs before heading home for breakfast.

By mid-morning I was back out in the fields, this time with Fern. As I'd mulled over the coming year the previous day, I'd made a decision to step up Fern's training, with a mind to running her in sheepdog trials at some point during the next twelve months. It was as close as I'd get to a New Year's resolution. Today marked the beginning of a concerted effort to bring her on as a working dog.

For a brief time in my shepherding life, I'd imagined training was as easy as falling off a log. When I'd set out to train my first dog Kim, my sole source of guidance was an eighteen-page booklet with line drawings and simple instructions. It was one of those old-fashioned publications that promised that if you did X then the dog would do Y, and sure enough everything it said was going to happen did happen. The only major error was committed by me rather than Kim.

A dog is generally controlled with a few basic commands: stop, start, go left, go right and 'that'll do' – the universal instruction for it to return to its handler. There are more advanced commands, for instance 'look back', which instructs the dog to leave the sheep it's working and go in search of another group. This is generally considered to be the most difficult instruction for a dog to master because most want to work only the sheep in front of them. As a result, it tends not to be taught until dogs are fully mature.

The four basic commands can be delivered verbally or, when working over greater distances or in bad weather when the voice is inaudible, via a whistle. The whistled commands vary according to the shepherd and include variations to speed the dog up, usually a hard blast of the existing command, or to slow it down, usually a quietening of the tone. The verbal commands used by shepherds in this country, on the other hand, are almost universal. Start is usually 'move on', while stop is either 'stop', 'stand' or 'lay down'. 'Away' is almost always used to send the dog off to the right, 'come bye' to set the dog off to the left. When training Kim, however, I could never remember which way round the left and right signals were. Working on the assumption that 'bye' was more similar in tone to 'right' than 'away', I decided to train Kim accordingly. So for her, 'come bye' was go right, and 'away' was go left. It's a practice I've never succeeded in breaking so I am still one of the few shepherds who operates with back-to-front commands.

Twenty years on, I now realised how lucky I had been back then. Kim was such a good dog she effectively taught me. If it hadn't been for her I might never have understood how much easier a well-trained dog can make a shepherd's life. Training had rarely been so straightforward since then, however.

Even the best and most naturally gifted dogs could be a test and none of my dogs had proved more of a challenge to train than Swift. Watching Ernie these days reminded me of his mother in a way that was almost eerie. Like Ernie, Swift was absolutely obsessed with sheep from an early age. From the time she was four or five months she had habitually headed off and rounded up sizeable groups of sheep at every opportunity. The problem was, she did it without any commands and I had no hope of stopping or controlling her. Some evenings, just as with Ernie, it had taken me half an hour to capture her. It had taken Swift months to decide to work with rather than against me. I hoped Fern was going to prove more straightforward. Derek had sent her down from Cumbria in the cab of a sheep lorry, when she was ten weeks old. In true collie style she had shown an interest in working within a couple of weeks, diving under the gates in the yards, startling the sheep before racing off.

I usually wait for the dogs to get to six months old before any serious training begins. Up until then Debbie and the children are usually in charge, taking them for walks and playing with them around the house. During her early days with the family, Fern had developed a liking for Laura's collection of cuddly toys, frequently raiding her bedroom and running off down the garden with one of her most cherished possessions.

Fern's qualities were plain to see. Her dainty size was built for speed while her large, wolf-like prick ears undoubtedly gave her an advantage when it came to hearing. There was no question she was potentially an exceptional dog. As I began her training, my job had been to mould these natural abilities.

The principles of training are based on the sheepdog's basic instincts, traits that have their roots in their ancient hunting ancestors. The impulse to chase livestock is not unique to collies, of course. As many domestic pet owners know too well, the sight of running sheep is almost guaranteed to stir the hunting gene in almost any dog. What is unique to the collie, however, is its herding instinct, the way in which by using its intelligence it has learned to circle and hold a flock of sheep together.

The modern collie's ancestors would have worked as a team, holding their prey together before cornering and killing it in a far more efficient manner than other canine breeds. Underpinning that team work, however, were the hierarchical principles of pack behaviour, and in particular the status of the dominant male. The leader would, for instance, always be the first to eat. It is this principle that lies at the root of working and training a modern sheepdog. My task, initially, is to establish myself as the pack leader so that the dog respects me and is willing to accept my commands.

It's usually reckoned that a dog is at its best when it has as many years behind it as it has legs and tail, that is to say five years old. But by the time a dog is two it should be fairly good under command and should know the basics. Fern at eighteen months had already done well, certainly in comparison with her troublesome kennelmate Ernie.

When I first took Fern to sheep at eighteen weeks old she was far too young, and it was really for my own curiosity to see how she would react. On that occasion, at the sight of the young pup running towards them the reluctant 'training assistants' bolted for the bottom end of the field, with Fern in hot pursuit. She eventually caught up with them as they were halted by a closed gate. Undeterred, Fern grabbed a mouthful of wool on the back leg of the nearest sheep and hung on, as she was dragged by the startled animal back up the hill. I eventually caught her and returned her to her kennel. She'd done no harm to the sheep, other than to pull out a little wool, but it was another six weeks before I had given her another chance to show me she was ready for training.

By then Fern was six months old and her young mind was developing from that of a playful pup to that of a slightly more thoughtful dog. When I let her run loose this time I could see the beginnings of that all-important herding instinct. She tried as best she could to keep the sheep together and then to bring them roughly towards me – a sure sign that she was ready to start her training for real. Since then I'd spent fifteen minutes most days training her.

Training a sheepdog takes many years of hard work and requires a different approach with each dog. Yet the starting point is always the same: the dog must learn to 'balance' or hold the sheep in position as inevitably they try to run off. Many dogs will do this instinctively – it is part of its hard-wiring to prevent sheep from escaping. My job is to use this knowledge to almost trick the dog into making the movements I need. If, for instance, a dog is on the far side of a flock of sheep and they begin to move to my right, the dog's instinct is going to send it running to its left to head them off. By my anticipating this and giving a distinctive command before it moves, the dog will begin to associate this instruction with running in this direction. By the time this and the other key commands – right, stop and move on – have been repeated over a period of weeks and months, the association between these unique words and whistles and the dog's instinctive movements will hopefully have become ingrained.

Fern's progress had been quite rapid. She had grasped the art of 'balancing' sheep fairly quickly and was responding well to all my commands. But it was her quickness of movement that really made her stand out for me. Good sheepdogs control sheep with subtle, sometimes tiny, movements. In training, Fern never took her eyes from the sheep, and was soon reading and responding to their every movement, with a lightning speed and a real economy of effort. If I could put it all together I was confident that I had a really good working and trialling dog in the making.

If Fern had an Achilles heel it lay in her occasional lack of confidence. It was something I'd been aware of for a long time, and I was always anxious not to put any pressure on her. In fact I almost treated her with kid gloves. I'd tried to remedy this around the farm, primarily by giving her easy jobs that she could succeed in doing and using gentle commands. There seemed no other answer than the long-term approach; I couldn't simply train her to be confident.

That afternoon on the twenty-acre field above the farm, I decided to set her off to gather a small flock three or four hundred yards away. To reach them she had to run through a gap in the earth bank. It wasn't a difficult 'outrun' but neither was it completely straightforward.

Immediately I could see the familiar signs of her lack of confidence. Fern ran ten yards then stopped, looking back at me and wagging her tail slightly. In her mind she was clearly not sure what I wanted her to do, even though the flock were clearly visible from where she stood. I gave her another command to move on but this time she lay down, ears tucked back, looking away. It was as if she was worried about what she was doing – or not doing. Eventually, I got her to get up and run off. For a moment I was encouraged; she was heading in the right direction at least. She had gone a hundred yards before things had gone wrong again. She had spotted a gate on the right that eventually led to our neighbour's garden. She disappeared under it and was soon lost from view altogether. I groaned to myself. For the moment, the thought of her running successfully on the trials field seemed a distant dream.

Two weeks into January and the winter landscape was almost drained of colour. The woodland stood lifeless, a dull grey, the grass was an ochre yellow. With the older members of the flock housed, even the white dots of sheep were sporadic in the fields. The winter routine was now under way in earnest. Every sheep on the farm was being fed both silage and pellets. In addition those in the shed needed bedding with straw every few days. The ritual was taking up four hours each morning – and another hour and a half each evening.

The sight of a familiar face climbing out of his pick-up truck this morning promised a welcome break from the monotony, at least.

At this time of the year, Michael was one of the busiest men around, travelling throughout Devon, Cornwall and Somerset scanning flocks of sheep midway through their pregnancies. Scanning is a relatively recent addition to the sheepfarmer's armoury, first introduced fifteen or twenty years ago. Although it's a comparatively expensive process, working out at about fifty pence a sheep, it does have huge benefits.

As lambing looms closer during the late winter, it's hugely important that you give each pregnant ewe the correct amount of food it needs. Whereas a ewe carrying a single lamb can almost look after itself for all but the last few weeks of a pregnancy, ewes carrying two or three lambs require a large amount of feeding and looking after in order that they can deliver viable, healthy lambs and provide sufficient milk for them to survive. Scanning tells the shepherd how many lambs each individual ewe is carrying, or, in some cases, if the sheep is not carrying lambs at all.

Michael was a man by whom you could set your clock. Punctual as ever, he'd arrived at nine o'clock, just as I brought the last of the four flocks in from their wintering fields. I'd started early but the job had taken longer than I'd anticipated – mid-pregnancy is no time to rush the ladies. I spent a moment or two trying to convince Michael that it was in fact he who was early. By the time he had set up his equipment in the sheds, I was just about organised as well, with the first of the girls ready to run through the race, Swift and Greg ready to force the sheep as required.

When Michael started his business fifteen years earlier, it must have been a brave decision to invest in what is fairly expensive, high-tech ultrasound equipment. Back then many sheep farmers, myself included, felt scanners were unnecessary and would never catch on. Michael's gamble had paid off handsomely, however. On his annual rounds, he now scanned 70,000 sheep.

Despite having what was obviously a thriving business, Michael continued to sit on the same orange plastic chair he'd had when he first came to the farm. Broken and taped-up, it had one leg shorter than the other and needed propping up. It must have seen the inside of a thousand lambing sheds by now. 'The money you're earning, I'd have thought you'd have treated yourself to a new chair by now Michael,' I said.

'Can't afford that,' he replied, flashing me a smile while he stamped his feet and rubbed his hands in an effort to warm up before the marathon ahead.

As ever, I was amazed at the speed with which Michael put the sheep through. To the layman the screen of his monitor was always a mystery. However hard I stared at the television-style display, I always found it difficult to find any clues to how many lambs each ewe was carrying. I could occasionally make out what he assured me was the spine of a lamb, but little beyond that. Michael, on the other hand, could put the sheep into a head clamp, run the scanning head under the belly, count the lambs and let it go in a little under fifteen seconds. His speed and accuracy were such that I'd never known him to make more than three or four inaccurate calls in a year. Once more he just sat there barking 'triple', 'double' or 'single'. Debbie had arrived back after the school run and was in charge of putting a spray mark on the ewes' shoulders – orange for triplets, blue for twins and green for the singles. But at times it was hard to do even that, such was the speed that sheep were passing through.

By one o'clock Michael had scanned all five hundred sheep. Debbie brought out some coffee and we did a quick tally. Only a dozen ewes were barren. By Michael's count there were 320 doubles, 45 triplets and 123 singles. This meant I should be expecting a total of just under 900 lambs.

Debbie asked Michael whether he'd like to stay for lunch, but he was already draining his coffee. He was soon heading back up the lane on his way to another farm, still on schedule. Only one thing, it seemed, was going to interfere with his immaculate timekeeping – and that was the total collapse of his chair leg.

CHAPTER NINE

Gate Expectations

Time has a habit of disappearing at an alarming rate between the New Year and lambing. What seemed like only a day or two after the scanning, I'd squinted at the calendar and found a fortnight had flown by and there were now just six weeks before the first arrivals were due. Time for the flock to be vaccinated.

Newborn lambs are prone to a range of potentially fatal diseases. The most common are the clostridial diseases, blackleg, braxie and lamb dysentery, and pasturella, a highly contagious bug that can cause either pneumonia or septicaemia. Another common threat is Orf, locally known as lore, a nasty problem that can affect young lambs as young as two weeks old. Orf has two forms, benign or malignant. It starts with small lesions or warts that appear on lambs' gums, then spread up out of the mouth towards the nose. In the worst cases these lesions become massive, almost like a large bunch of red grapes around the mouth, then spreading into the mouth, the gums and the tongue. It's also highly contagious and can pass on to the teats of the ewe from which the lamb is suckling, with the result that the ewe too is infected. The ewe then resists any attempt by the lamb to get milk and the lamb starves. The result is an emaciated lamb that often dies because it can't survive on grass at such a tender age.

The best way to avoid each of these is through vaccination. For pasturella and the clostridial diseases there is a combined 'seven in one' vaccine that can be administered in one dose. Orf has its own specific vaccine. In both cases, immunity can then be passed on through the first milk at lambing time.

The sun had made a rare appearance as I walked across the yard with the metallic box in which I keep my medical equipment. I let Greg and Swift out to lend a hand in the sheds. The scanning had allowed me to divide the ewes up for the final phase of their pregnancies. I can house at most 450 sheep under cover at Borough Farm, so I'd had to prioritise. I put those ewes carrying triplets in one pen, the early lambing ewes in another two pens, with the balance of the pens being made up by ewes in poorer condition and those carrying doubles. Two weeks after the ewes had been moved under cover they had taken to their new environment so well that getting them out of their pens had become a real challenge.

As George Orwell almost said in *Animal Farm*, all sheep are equally stubborn but some are more equal than others. This morning they were again resisting any attempt to force them out of the gates. One of the great frustrations of sheep is that they appear sometimes to know exactly where they should be going but do everything possible to resist. Today the entire pen seemed to have decided to refuse en masse to go out through an eight-foot gate despite the fact it was wide open. Fed up with trying to force the issue, I'd left the gate open and hidden around the corner for a moment. You invariably find that within a minute a sheep spots what it takes to be a gate that has been accidentally left open and charges out of it. If you then follow up the last sheep you can get them out without any trouble. Sure enough, it worked again. No sooner had I sent the dogs elsewhere and gone round the corner than they were moving in the direction I needed. Another impressive insight into the intellect of the sheep!

Vaccinating a sheep is never a straightforward job. It is not like a doctor giving an injection to a compliant patient. Sheep tend to react rather differently, usually leaping up on to their back feet and jerking uncontrollably in the air whenever a needle is inserted in their neck. The greatest danger here is that you end up sticking the needle into yourself. I've often had a sheep rear up while injecting it, sending the needle into my thumb. Strictly speaking, when this happens I should follow the instructions on the label that say 'Seek Immediate Medical Attention'. You don't go, of course, although the after-effects of the injection are quite painful for a day or so.

If the labels are to be believed, the most dangerous injections are those for footrot and Orf. You could lose your finger by accidentally injecting yourself with the former – which is one of the reasons why I don't use it. With the Orf vaccination, you can inadvertently give yourself a dose of the disease – an unpleasant experience I've been through once and am in no hurry to repeat if I can help it.

Administering the Orf vaccination is, I have to admit, my least favourite job of the year. It's given by means of a device that's mounted on the end of a bottle, and is delivered by scratching across a bare piece of skin on the ewe's anatomy, usually under the tail. It dispenses a live virus that triggers the immune system of the ewe, which then passes on to the lamb when it's born. Trying to deliver the vaccination to the underside of the tail of a moving ewe can be very tricky, particularly as the sheep's waxy skin tends to clog up the needle.

This morning, at least, the sheep behaved no worse than normal. I had long ago learned how to jump out of the way when a ewe bucked up in the air after receiving her injection. I'd developed a technique for pulling the syringe out of the way to avoid jabbing myself. By lunch time I'd got through two-thirds of the flock. By late afternoon all five hundred were safely vaccinated. There weren't many occasions during the shepherding year when I was glad to see the back of a day – but that was one of them.

The wind was already howling and the rain driving in so hard it crackled on the window panes in the kitchen. According to the early morning weather forecast on the radio, things were going to get even worse as the day wore on. 'I must be completely mad,' I said to myself as I opened the door and leaned into the gale.

It was the final Saturday of January and I'd decided to take Fern to a nursery sheepdog trial an hour and a half's drive away, down in Cornwall. Nursery trials were relatively uncharted waters for me. They were an excellent idea, designed for young novice dogs to get a taste of competition in a new, unfamiliar environment. This one was being held at the farm of a well-known figure in the sheepdog world in the South West.

As I headed along the north coast, the Land Rover's lack of third gear made driving into the headwind even more difficult. Each time an occasional gust threatened to push us off the road, I found myself wondering why I was doing this with a dog about which I was unsure. It really did feel like I was doing the most ridiculous thing. Yet at the back of every sheepdog triallist's mind there is a ridiculous level of optimism. It is a little voice that says, 'This time you've got the perfect dog, the one that's going to win dozens of trials and go on to be an international champion.' And, despite the problem with her lack of confidence, this was the feeling I had in my mad moments of optimism about Fern.

In North Cornwall the weather was no better. As I arrived at the farm where the trial was to be held, a group of handlers huddled behind a dozen cars, their hoods turned up against nature's onslaught. I guessed there were familiar faces underneath them but, with the weather about as hard and relentless as it gets in this part of the world, there was no recognising anyone.

CHAPTER TEN

Storm Clouds

With Debbie and the children in bed and the house shrouded in darkness, I slipped out into the quiet cold of the February night and headed for the lambing sheds. I'd started doing late-night checks on the ewes a couple of weeks earlier, at the start of February. Lambing was just two weeks away now and, with the expectant mothers susceptible to a variety of problems, I tried to maintain as close to round-the-clock monitoring as I could manage.

When I'd begun these late-night vigils, the loud metallic clanging of the door as I entered the shed had startled the ewes, sending them scurrying nervously into a huddle in the corner of the pen. By now, though, my arrival caused barely a ripple of interest. The flock lay there, contentedly chewing the cud, a cloud of steaming breath rising above them.

The relative calm and quiet of the shed tonight meant the distinctive low bleating of a ewe cut through the night air even more sharply. This long before lambing, it was a sound that could mean only bad news.

Turning the beam of my torch in the direction of the bleating, I picked out a ewe in the corner of one of the pens, clearly distressed. A closer inspection confirmed my worst fears. In the half light I could see that the ewe's back end was bloodstained. Climbing into the pen, I soon found the distorted shape of an aborted lamb lying in a pool of dense brown blood in the corner. The distinctive dark red 'buttons' on the placenta indicated the lamb had become infected in the womb where it had died days before. It had taken until tonight for the ewe's body to expel it.

The sheep's distress was understandable. Her bleat was a low and maternal one, a sound that sheep produce only for the first few hours of motherhood. But she didn't know whether she'd produced a live lamb or not. Confused, she kept turning round and sniffing at the aborted lamb in the straw. Each time she did so she walked away, bleating more loudly, as if in disappointment.

My heart sank. Although a small proportion of the ewe flock will 'slip' their lambs each year through natural stresses and strains, the various forms of contagious abortion that can be passed from one ewe to another are a real danger. Here too, a sheep's worst enemy is another sheep, and in the confines of the sheep shed such an infection can easily be passed through contact with diseased material. When this happens, exposed ewes lose their lambs a week or ten days afterwards. In the worst case the problem can multiply out of control in a frighteningly short space of time. So my immediate priority was to try to prevent other ewes becoming infected.

I cleared up the lamb as best I could, putting it into a bag for disposal. I then brought the ewe out to separate her from the rest and put her in a small pen by herself. She looked comparatively bright, which was a good sign. Her ears were up and she showed some interest in the handful of silage I gave her.

In the worst cases of abortion, the infected foetus or foetuses start to break down, and are not expelled by the sheep. Perhaps the worst job for any shepherd or vet is manually removing such lambs, in their semi-rotted state. All too often septicaemia sets in and the ewe becomes gravely ill very quickly. Thankfully, judging by the appearance of this sheep, she would make a full recovery. I drew up some penicillin and injected it into her back leg, just as a precautionary measure.

Strangely, I had worried about this sort of situation far more as an employed shepherd in Kent. Any conscientious stockman takes a threat like this very seriously, but somehow, having to explain it to an employer who didn't necessarily understand the situation just added to the worry. In recent years I'd become a little more philosophical. Beyond isolating affected ewes and practising cleanliness, there was only so much you could do to prevent the spread of infection. The rest was down to Mother Nature, luck or whatever. Worrying about it didn't do any good. The problem was, worry is just part of human nature.

Fortunately, one abortion doesn't make a 'storm'. I just hoped this was one of those isolated cases that often crop up just before the start of lambing, and not a sign of something far more severe. It was gone one o'clock when I slipped into bed. I'd been through all the possible scenarios a dozen times in my head before I finally went off to sleep.

It doesn't do any good to worry on these things too much. So the morning after I'd discovered the 'slipped' lamb, I finished the rounds of feeding and bedding then took Ernie out to the small paddock behind the house where I had a dozen 'training' ewes. These older ewes were so used to the idea of being rounded up they'd become almost obliging to inexperienced youngsters. They were the ideal sheep on which to start young dogs working.

Ernie's progress could not have presented a starker contrast to Fern. While she had made steady strides in the past months, Ernie had remained a law unto himself. His obsession with sheep had, if anything, grown. Given any freedom, he would take off to the fields in search of some 'work'. By day, this meant that I had to restrict him with a long lead at all times. This was denying him the exercise he needed, so I had begun giving him a run after dark each night. Even this hadn't worked, however, and things had come to a head a few nights earlier.

I'd been ambling back from a walk with the dogs in the darkened woods and was returning through the lower fields when Ernie had once more disappeared. Straining my eyes in the blackness, I repeatedly called his name but to no avail. A few minutes later I spotted the silhouettes of the rams appearing out of the darkness. I knew full well who was driving them. As I walked up the hill, followed by all eighteen rams, keeping impeccably to my heel, I saw the occasional black shadow darting behind us, keeping the sheep in line.

Ernie drove the rams as far as the gate at the top of the fields, but even then my workaholic pup wasn't ready to return to me. After five futile minutes calling him, I resorted to rugby tackling him, quite a challenge in the dark. I managed it but, even with my fourteen stones lying on top of him, Ernie still managed to wriggle free in pursuit of the by now dispersing rams. I'd been left lying on the ground, where I slowly came to the realisation I'd landed in a rather ripe pile of sheep dung.

It was only exhaustion that had brought Ernie's exertions to an end. But as I'd finally clipped the lead on to his collar, I still couldn't bring myself to reprimand him. He was after all following his deeply ingrained genetic instinct, only doing what every nerve in his body told him to do. And he was hardly the first dog to display such an obsessive interest in sheep – his mother Swift had made the odd midnight excursion into the fields herself. All I needed to do was to get him to express his instincts under my control and at the time of my – not his – choosing. I had gritted my teeth and told him what a good boy he was. Inwardly, however, I had also resolved his training must start in earnest.

I'd had a soft spot for Ernie from the very beginning. He'd been one of a litter of eight pups born to Swift, but had immediately stood out. In a litter of pups there is often one dog who has more initiative than the others. I had seen it with Gail in her litter. In this case it was Ernie. Very soon he had learned how to jump out of his run. From there he learned how to get to the back door of the house, and would appear there a few times each day. It was quite endearing, but it also showed that he had a bit more personality and a bit more gumption than the rest.

That year, as usual, we had some vet students on the farm to help with the lambing. One of them spotted that he had an umbilical hernia, about the size of a fingernail, on his belly. 'He's got a hernia so we've called him Ernie,' they announced one day, seemingly pleased with themselves. I'd thought: 'That's ridiculous, you can't have a dog called Ernie.' Sheepdog names are usually one syllable, so they can be spoken quickly. Because the relationship between shepherd and sheepdog is a serious working relationship, they are also often the sort of names a workmate might have, names like Bob, Dave, Ted or Greg. But by the time the lambing students left he was answering to their chosen moniker, so that was it. I was stuck with it.

From an early age, Ernie's single-minded determination to work sheep really was remarkable. I began today by sitting down beside him and spending a few moments talking gently to him, trying to make eye contact with him, and looking for some sign that he might be willing to listen to me when we got down to some work. With some reservations, I then got to my feet and walked him, still on his lead, towards where the sheep were now grazing.

I slipped him off his lead and made a whooshing sound to encourage him to run – as if he needed it. If I could just get him to the other side of the sheep, keeping ten yards or so back from them and to stop, I would at least have made a start. For the next few hectic minutes I had to move quickly to stop Ernie from completely circling the ewes. I growled at him firmly every time he came too close.

To an onlooker the scene would have looked chaotic, but then came the moment I'd been waiting for. Ernie made a wide, sweeping arc and arrived at a point where he obviously felt he had his charges under control. But perhaps he also sensed that I was not going to let him have everything his own way. He was some distance from the sheep, the sheep were relaxing and not making any attempt to break away.

'Stop, Ernie,' I said in a firm voice. He moved forward slightly.

'Stop, Ernie. Stop.'

Without taking his attention from the sheep for a moment, he stood stationary for a full five seconds. Then a movement from one of the old ewes tempted him to be off once more, but it didn't matter. Ernie had listened to something that I had said to him while at work, a major victory in itself. A basic rule of training is to always finish on a good note, so I decided to call it a day. A few minutes later, he had even allowed me to catch him without doing an impersonation of Jonny Wilkinson or Lawrence Dallaglio once more.

'One small step for most dogs, one giant leap for Ernie's kind,' I thought to myself, shutting the kennel door behind him.

Two days after the first abortion, the morning's inspection of the flock revealed another bleating ewe. She didn't appear to be as distressed as the previous one. As I approached her the ewe seemed fine. But on looking closer I found the telltale sign of blood around the tail. A few moments later I found two aborted lambs barely visible in the deep straw.

With two aborted ewes in three days, it would have been easy at this stage to start fearing the worst.

Of the seven causes of contagious abortion in sheep, two are more common than the rest. The first, toxoplasmosis, is commonly spread through rats and then – even more virulently – by cats. It's said that the faeces from one infected cat can carry enough toxoplasmosis to make every sheep in the country abort. Strangely, once the cat becomes older its immunity is such that it doesn't present a threat.

My fear was that it was the more contagious of the two forms of abortion – enzootic, which has been on the rise in the last twenty years. The only way to be certain was to have a laboratory investigation carried out. I put some of the infected placenta into a plastic bag and headed for the vet's surgery.

The rural veterinary practice has changed out of all recognition in recent years. The days when James Herriot dropped by for a cup of tea and a chat have been consigned to history. Twenty years ago, farm-related work would have made up seventy per cent of the business for a country practice like the one I use. Now that figure would be less than twenty per cent, and the vast majority of that would be from dairy farms.

A sheep's welfare should never be compromised on the grounds of cost, of course, but the harsh fact is that in most cases the bill for a veterinary intervention is more than the value of the animal. So it's inevitable that the days when a vet would be called out to treat an individual sheep are almost gone. Nowadays vets tend to be called in to look at major flock problems or to advise on management of a farm's population of sheep.

So as a result local practices have geared themselves to deal with a very different type of client. At my local practice, the separate reception for farmers has long gone. The foyer and the whole atmosphere is orientated to small animals, dogs and cats in particular. I frequently received disapproving looks from the receptionist if I forgot to leave my dirty boots at the door.

As I walked through the waiting room today, it was a large Labrador that was being reprimanded. The dog's owner was telling him off for taking a keen interest in the plastic bag in my hand. I suspect he'd have been even less popular if his owner had known what was inside.

The receptionist on duty was Jackie, who specialised in looking after the practice's farming customers. Until recently she had kept a small flock of sheep herself. She understood the worry that a bout of contagious abortion could bring, and as I handed over the well-sealed specimen she was reassuring.

'It usually takes about four days for the results,' she said. 'I'll give you a call when there's any news.'

Nothing I could do but wait, I told myself as I headed for the door.

A week since the first abortion – and four days since I'd sent the sample to the vet – my anxiety levels had been ebbing and flowing. In the past three days I'd found two more ewes that had aborted. One had been in the sheds once more, but the other had been in the fields, which was significant, I thought. The fact that a ewe that had not been in contact with the shed had aborted probably meant that, if the problem was the result of an infection, it was one carried over from last year.

I'd also drawn comfort from the fact that we hadn't seen a great flurry of lost lambs. But things were still too unclear for me to relax. A phone call from Jackie did little to ease my concerns.

'There was some toxoplasmosis, but it may not have been the cause of the abortion,' she said, almost apologetic at the indecisive results. 'It could just mean the sample was too old when we received it.'

It wasn't uncommon for laboratory results to be unclear but, as I mulled over what Jackie had told me, I could see positives. To my relief, they had definitely ruled out enzootic abortion, which to my mind was potentially the biggest problem. Although toxoplasmosis can cause major 'storms', in my own experience it didn't often result in a significant outbreak.

I would still be on edge every time I went in to feed the sheep over the coming days. With a week to go to lambing, the numbers could still escalate from here. But if we could just hang on for a few more days, the arrival of the first few lambs would, I knew, make things seem much better.

CHAPTER ELEVEN

The Night Watch

Deep in the darkness of the woods beneath us, the screeching of an owl cut through the night. 'We might be in luck,' I whispered to Clare as we made our way through the lower fields, my torch picking out the path ahead of us.

The owls had been particularly vocal in recent nights, so Clare had jumped at the chance of coming out on one of our occasional night-time nature walks. Laura had been disappointed at having to stay at home. 'When you're a little older,' I'd consoled her.

The conditions were ideal with plenty of cloud cover, and enough wind to mask our scent and noise. Clare was protected from the chill easterlies with a balaclava, gloves and at least three layers of jumpers.

On our most successful nocturnal excursion a year or so earlier, we had switched on the torch and almost immediately chanced upon two Little Owls on the lower branches of an overhanging ash. They had both sat there, peering towards the light for a minute or so before disappearing into the depths of the woods. We often heard them, but it was a rare treat to see them. Sadly, we had no such luck tonight. As we scoured the trees at the woodland edge, the darting beam of my torch on the lower branches of the ash picked out nothing more than the fleeting glimpse of a wood-pigeon, flapping as it spiralled away into the pitch black.

'Let's see if we have any luck at the badger sett,' I told Clare, taking her hand and pressing on to the opposite side of the valley.

With Clare clutching my hand tight, we followed the edge of the woods, around through a shallow cleave, where we both clambered over the fence and into the woods. The farm was home to one sett, but the spot the badgers had chosen was one of the most inaccessible on the farm, a steeply sloping bank covered with vicious brambles. Although they were dormant for winter, the brambles were still capable of tripping and scratching us as we scrambled through in the darkness.

After pushing our way through the undergrowth for a few minutes we located what seemed to be an active entrance to the sett. My torch picked out what looked like fresh markings and signs of excavation. We found a nearby log and sat down. 'We'll give it twenty minutes,' I whispered to Clare. 'See if there's any sign of him.'

Clare didn't complain about the cold, or seem to be fussed about sitting in the pitch black on a damp log in the woods. Although I didn't hold out much hope that he'd put in an appearance with us sitting so close by, I didn't rule it out completely. Several years earlier while out walking late, I'd watched a pair of badgers for several minutes, sniffing out slugs and insects within a couple of yards of my feet, oblivious to my presence.

There were several false alarms when we both thought we heard movement in the nearby scrub. But each time we were disappointed. After twenty minutes I could hear Clare's teeth beginning to chatter in the cold. We scrambled down the steep slope, crossed the stream and started heading back towards the farm.

Clare's disappointment was obvious. 'All we've seen in the last hour and a half is some not very rare sheep!' she said, a little grumpily. Fortunately, the final climb up to the farm transformed her mood.

As we neared the field gate at the edge of the woods, I was aware of movement about fifty yards away from us. I swung the torch around. For a moment a young fox, probably one of last year's cubs, stood frozen in the beam of light, the limp form of a rabbit visible in its mouth. The fox stared back for a second or two before darting off back into the darkness.

'Did you see that, Dad?' Clare said, clearly impressed at the sight of one of nature's dramas unfolding in front of her. She was still chatting animatedly about it as she climbed into bed half an hour later.

With the house asleep, I slipped out into the cold once more to check the ewes in the sheds. I knew what I was going to find almost as soon as I opened the door. A ewe was bleating once more, but this was a gentler burble and was accompanied by the echo of other, fainter cries. Already on their feet, a pair of lambs were steaming slightly as they shivered in the evening cold. A delighted ewe was licking one then the other in turn, nuzzling them gently in the direction of her udder as she did so.

To hear that maternal bleating for the first time in the year is always something a bit special. Given the concerns I'd had over the abortion storm, the arrival of the first lambs sounded even more magical than usual.

As ever, I wasn't organised for the onset of lambing. Using some hurdles, I hastily made a small pen, four foot square, then bedded it deeply with fresh straw. I returned to the ewe and picked up the two lambs, holding them by the front legs in front of the ewe's nose. She was an older ewe and followed them easily to the pen.

Both lambs were a good size, six pounds. They'd have a good chance of survival at that weight. They – and their mother – needed a couple of standard treatments, nevertheless. The unsealed umbilical cord provides an easy route for dangerous bacteria to invade the lamb's body, so I applied a strong iodine solution to their navels. A good soaking with the solution would shrivel and dry the cord over the next twelve hours. I then checked the ewe's teats.

A newborn lamb is totally dependent on its mother's warm first milk, colostrum. Yellow and almost as thick as cream, it is rich in antibodies that provide the lamb with immunity to early diseases. More importantly, for the first few hours it also provides the energy that allows the fragile young body to generate warmth. A lamb that doesn't receive enough colostrum during the first twenty-four hours after birth is almost certain to die.

Leaning gently against the ewe, I checked her udder. It was full, but often the end of the teat can have a small waxy blockage, which makes it impossible for the milk to flow. I squeezed a jet of milk from each side then sat back, hopeful that the miracle of nature would now unfold.

Every shepherd must feel the same. And it doesn't matter how often I see it. Those first lambs of the year are something to be savoured. To watch how, within a few minutes of their birth, through trial, error and some amazing instinct of survival, they've managed to find their mother's milk and suckle is a thing of wonder. In the darkness tonight, I witnessed it once more. Within five minutes each lamb had found the teat and, judging by the soon rounded bellies, had taken a half pint of milk from their mother's full-to-bursting udder.

As her lambs fed, the ewe almost murmured contentment. Her maternal instincts were strong and the undivided attention she was showing her offspring would greatly increase their chances of survival. Turning to each youngster in turn, she once more licked away the birth fluids, drying the lambs to keep them from chilling.

I usually leave each new family unit in their pen for forty-eight hours, long enough for the maternal bonds to strengthen. During the next two days I would check the lambs every few hours to make sure they were disease-free, feeding well and that their ewe hadn't rejected them. I could see there was little chance of that with this ewe. The first lambs of the year were in good hands.

Satisfied, I hastily tied up half a dozen hurdles to make three impromptu pens in case there were more arrivals during the night. As I did so, the anxiety of the last three weeks was already lifting. I'd had no more than half a dozen abortions, less than average on many farms. The nightmare of this turning into an abortion 'storm' was receding.

The novelty of spending evenings here in the sheds rather than inside the house would soon fade too. Lambing – and its myriad complications – would become a seemingly endless challenge. For now, however, I couldn't resist enjoying the first lambs of the year. Why not? I couldn't imagine I would be able to spend as much time with any of the thousand or so others that would arrive in the coming weeks. I turned up my collar against the now freezing night air, and watched them for ten minutes before finally turning in for the night.

It was well past midnight as I wandered back to the house. The night was clear, crisp and starlit. There was no moon but the constellations were shining brightly as I gazed upwards.

I often wondered whether on some distant planet there was someone else daft enough to stay up half the night looking after a bunch of sheep. On nights like this, however, it didn't seem so daft after all.

Spring

CHAPTER TWELVE

A Family Affair

Two days on from the first, midnight delivery, and the lambing had yet to gather any real momentum. As was already the norm, the alarm had summoned me from my bed at three o'clock this morning. In total only twenty ewes had produced so far, and in the darkness today I'd found just one new lamb. It had given me the chance to grab another three hours' sleep and now, as I made my way across the yard once more – the steam from a mug of hot tea rising into the cold of the early March morning air – I sensed I'd soon be grateful for it. Yesterday had been the day the first group of ewes were 'officially' due to deliver, and sheep were nothing if not predictable. The rush had to start soon and sure enough, as I perched my mug on the end of a trough and looked around, the pens had become alive with activity during my brief absence. The organised chaos of the main lambing was upon us.

In the pen furthest from the door, three ewes fussed over three newborn lambs. Close by, a fourth lamb wandered in search of its mother, bleating among the other sixty ewes in the group. In the pen next door, things seemed in rather better order, with one ewe tending to a strong-looking pair of Suffolk lambs, while another lay contentedly by the side of what was obviously a single lamb.

Of the ewes that had lambed in the past hours, it was the one I spotted in a pen near the doorway that most needed my attention. She was fidgeting, turning around and around, scratching in the straw as if trying to make a bed. As I watched her for a second she lay down and threw her head in the air, straining. The reason for her distress was soon obvious. The correct presentation for a lamb entering the world is for the toes of each front foot to appear almost simultaneously with the nose. This allows it to pass through the gap in the ewe's pelvis with the minimum distress to mother or offspring. Frequently, however, the head emerges first leaving the legs trailing behind. This is a particular problem with larger lambs, whose shoulders become stuck, wedged in behind the ewe's pelvis, making it impossible for the unborn lamb to pass through. This is what had happened here – the ewe's lamb was arriving 'head out'.

This is a dangerous situation and needs to be dealt with quickly. As the lamb's neck is constricted, its head rapidly becomes swollen, often with fatal results. Fortunately, judging from the lack of swelling here, the 'head out' had only just happened.

Ovine obstetrics become second nature to a shepherd. Out of sheer necessity most are highly skilled at resolving the many tricky deliveries that will inevitably occur during the season. Grasping her back leg, I laid the ewe on to her side. Pulling on a disposable glove, I then inserted my fingers past the lamb's neck. You can find lambs in the most awkward positions, but fortunately this wasn't too bad. Feeling a knee joint, I hooked it forward and with this leg now correctly extended I pulled gently on it, while using my gloved hand to ease the lamb's other shoulder past its mother's pelvis. Suddenly it was free: the lamb slid out on to the straw bedding. The ewe thrashed, trying to get up, but I held her down until I could be sure that the lamb was breathing.

I wiped the mucus from the lamb's mouth and it gave a gasp. Usually nature would now take over and its chest would begin to rise and fall rhythmically. But the lamb remained lifeless, making no further attempt to breathe. I gave it a couple of slaps, which brought another, almost half-hearted, gasp, but no regular breathing pattern. I took a dry piece of straw and pushed it into the lamb's nose. This is an age-old trick which usually produces immediate results. It didn't fail me. In a moment there was a sneeze, and the lamb shook its head. Its rib cage pumped vigorously as air filled its lungs for the first time. A few seconds later and the lamb's head was up, the surest sign that its life no longer hung in the balance. I let the ewe on to her feet and backed away rapidly. The mark from the scanning said that she should be expecting a second lamb but, despite her difficulties, there was no reason why she shouldn't have the next one on her own.

This group of ewes was the one that had been run with the teaser rams back in the autumn. Ewes stimulated in this way tend to produce in one great rush, so

This is the sort of job that eats up the hours. Glancing at my watch, I realised I had spent nearly fifteen minutes with just two lambs.

A smaller lamb in another pen represented a slightly different problem. On checking its mother's udder, I found one side hard and yielding no milk at all – a clear case of mastitis from last year that wouldn't now be treatable. This lamb, one of a pair, had been born only the night before, and so was in no immediate danger. But its sibling was much livelier, and had obviously discovered the only side of the udder that was functioning.

Reaching for the marking box, I grabbed the most important tool in the lambing shed, the stomach tube – a large syringe, with a twelve-inch-long tube that passes milk straight into the lamb. I took the unfed lamb and held it upright between my knees. Using milk from the jug I'd drawn off a few minutes earlier, I fed the tube its full length into the lamb's stomach then squeezed gently on the syringe. It's a remarkably quick operation. In a short time, I'd delivered two full syringes directly into the lamb, ensuring that even if it didn't find any milk from its mother it would be safe for the next six hours.

By now Debbie was emptying the umpteenth water bucket of the morning. Laura followed behind refilling with a hose, while Clare fed an armful of silage to the last hungry sheep. My mother had also arrived, and was looking to help while entertaining Nick at the same time.

'Can we have a cup of coffee now?' Laura asked, perching herself on the old sofa that had been installed in the shed for the duration of lambing.

'Are you sure you deserve one?' I said with a smile.

CHAPTER THIRTEEN

A Fateful Choice

The continuous winter cycle, with its waves of strong winds and heavy rain blowing in off the Atlantic, seemed to have been broken at last – and not a moment too soon. The past few days had been fairly hectic, with thirty or more ewes lambing each day. By now every inch of available space for ewes and lambs was taken inside the sheds. I needed to begin turning some of them out into the fields.

All being well, it takes only a couple of days for the maternal bond between a ewe and her lambs to become fully formed. After that, assuming the lambs are strong, the ewe is milking well and the weather is favourable, the grazing and fresh air provide a far healthier environment for the family than the cramped sheds.

So with the skies clear and promising to stay that way, I selected some of the oldest, strongest lambs and carried them out into the morning air. Their mothers following behind, we walked across the yard and through the gate into the lower field bordering the woods – or 'Steep Sheep' as the children had named it.

damage looked severe. There was blood on Ernie's lips and on the leg of the ewe. The ewe was still standing her ground but now with a scared look in her eyes, her ears twitching, her nostrils sniffing in the direction of the danger.

I'd never had a dog cause damage to a sheep. They were there to assist me in looking after the sheep, not maim them. I was devastated – and furious. The anger in my voice as I boomed at him sent Ernie cowering to the floor. Still struggling to contain my fury, I grabbed him then dragged him back to his kennel, slamming the door shut. Of all my dogs this was the one I'd thought was a cut above the rest. I'd placed my faith in him and he'd let me down. How could he have done it? Why did he do it? For now, however, the sheep was my main priority.

Back in the shed I inspected the ewe properly for the first time. Although she was holding the leg off the ground while standing, she did manage to put it down tentatively as she walked. She was probably going to be OK. Straight after feeding the rest of the flock, I cleaned and dressed the wound, treated the ewe with antibiotic, then wandered miserably back towards the house. I couldn't even bring myself to look in the direction of the kennels.

Debbie immediately sensed something was wrong.

'What is it?' she asked.

'Ernie,' I sighed. 'He's attacked a ewe.'

I could feel a tear welling in my eyes as I recounted the events of the morning.

'What are you going to do?' Debbie asked when I'd finished.

'You know what I *should* do,' I said, looking up for a moment to catch her eye.

Debbie didn't respond. She just looked away.

There was work piling up outside, but for the moment I couldn't find the enthusiasm to return to it. As I sat in the kitchen, staring into my coffee mug, my mind was still racing, the same question hanging there: was it he who had let me down, or I him? In truth, I knew the answer.

The line between hunter and herder is a thin one – and Ernie still didn't know where it lay. For the first time in his young life he'd been faced by a ewe that hadn't turned and run from him. He'd had no other instinct than to stand his ground. The blame lay with me. I shouldn't have allowed him to get into a situation which he didn't have the maturity to understand, or the experience to handle.

By the time I drained the last remnants of cold coffee from my mug, however, the question of what had happened and why had been replaced by an even unhappier one. When to make the dreaded call to the vet? On many farms, there wouldn't have been any hesitation. Many farmers would have concluded this was a dog that was simply too dangerous to have around sheep.

I couldn't face it now. I felt too overwhelmed with the emotion of the moment. I needed to think about it in a cooler, more detached frame of mind. I left the house and wandered back across the yard once more. As I walked past him, Ernie sat at the front of his kennel, chest puffed out as proudly as ever, not a hint of shame in the faithful eyes that followed me.

CHAPTER FOURTEEN

Helping Hands

By the third week of March the first phase of lambing had already drawn to a successful conclusion. There had been few problems and, despite the fact that the first group contained the oldest members of the flock, the lamb tally appeared to be quite respectable. I say 'appeared', because I never like to count the lambs until they are sold. To do so earlier only seems to be tempting fate.

Over the coming days I would turn out the remaining lambs to the fields, clean out the lambing pens and generally prepare for the start of round two of the lambing. I would also try to ensure I got a few good nights' sleep. Even in my exhausted state, I found it hard to settle most nights. Ernie – and the issue of what to do about him – was still playing heavily on my mind.

Thankfully the damage to the ewe he'd attacked wasn't now looking too bad. It's a curious paradox that sheep can be dragged down very quickly by a relatively minor illness, but seem to be able to withstand some quite severe physical injury.

The ewe's recovery somehow gave me an excuse to keep postponing a decision on Ernie. So too did the arrival of this year's vet students.

Katherine and Amy were both from Cornwall and were bright, effervescent personalities. Laura and Clare were particularly excited at the arrival of two teenage girls on the farm. I showed them around the lambing shed, taking them through the basics – everything from penning to navel dressing, how to use the 'rubber rings' to castrate the male lambs and remove each lamb's tail. I then explained, as best I could, our system of spraying different coloured numbers on to each of the lambs so as to match those marked on their mothers. 'Blue for a pair, red for a single,' I said, already detecting a glazed look on their faces.

Being vet students, they had a good grounding in the various ovine diseases that were likely to crop up. Both seemed aware that the biggest problem they were likely to encounter was watery mouth, a common disease in housed sheep flocks, caused by a bacteria that the lambs ingest soon after birth. It's a disease that's easy enough to spot. The first sign is usually a wet mouth followed by drooling and a loss of appetite. Prompt treatment clears it very quickly, but if a lamb is missed and left untreated for a few hours its chances of survival diminish rapidly.

Katherine and Amy's timing was perfect. Within a day or so of their arrival the main flock began delivering. Almost immediately I got the feeling both girls had a natural aptitude for the job. Katherine's father kept a few sheep, and Amy had spent time working with livestock in the years before university. From a distance I watched them, ably assisted and supervised by Clare and Laura.

If there is one area in which Clare excels it's fostering lambs. Every year produces a few lambs who, for a variety of reasons, will be unable to get sufficient milk from their real mothers, but the most common 'fosters' are triplets. Because sheep are equipped with only two teats, they rarely manage to rear more than two lambs successfully. This year, forty or so triplets shown up during scanning were going to need fostering on to a ewe with only one lamb. Clare soon had an opportunity to pass on her technique to the newly arrived students.

She spotted a ewe marked as carrying a single scratching at the straw bedding, turning around, obviously in labour. The most successful fosters are achieved if the ewe doesn't first see her own lamb. So under Clare's instruction Amy caught hold of the ewe and held her so that she couldn't see her lamb as it arrived. At the same time Clare arrived with a bucket into which, with a little help from Katherine, she collected the embryonic fluid covering the lamb. It is this fluid which contains the scent that the mother uses for identification. Their task now was to use the fluid in the bucket to wash both the newborn lamb and a foster lamb so as to (hopefully) convince the ewe she'd given birth to twins.

I selected a strong, well-fed triplet from one of the small pens, and passed it to Clare for its bath.

'You have to make sure it's well soaked,' she said to Amy and Katherine, who were by now looking a little surprised at being taught by a nine-year-old. 'Otherwise the ewe will know it's not hers.'

She was right – although some ewes are fooled straightaway, others can detect the impostor at the first sniff. In these cases I resort to desperation measures, and add a little warm water and a measure of Dettol to the bath. The smell of the Dettol is so strong that the ewe, unable to tell one lamb from the other, usually resigns herself to life as a mother of two.

Clare finished washing the two lambs together, but then laid only the foster lamb at the nose of its new mother. The ewe sniffed it a little then gave it a tentative lick. For a moment the lamb shivered and nuzzled closer to the ewe. For a few seconds the ewe looked around as if to make sure there wasn't another lamb that could be hers. It took a little while before she had turned her full maternal attention on to the foster lamb.

'Looks as if she'll be OK with that one. Well done, Clare,' I said. 'Give her her own lamb in a couple of minutes.'

Clare wasn't totally committed to fostering all needy lambs, however. By the end of lambing, she and Laura always aim to have a pen of bottle lambs of their own to feed. This isn't something I agree with. To me, bottle-fed or 'tame' lambs are incredibly time-consuming. In addition, their powdered milk is very expensive, and they rarely thrive as well as naturally fed lambs. Unlike the girls, I consider it a victory if I can get to the end of lambing without any 'tame' lambs.

Katherine and Amy soon settled into the lambing routine. Their youthful energy saw them through the long hours and sleep-deprived nights and they seemed to be rather enjoying the work. It wasn't long before I felt comfortable leaving them to run the lambing shed for much of the day. They soon became dab hands at fostering and I was pleased to find that most of the triplet lambs were finding their way on to new mothers.

'Looks like you've managed to foster all the spare lambs away,' I said to Amy one evening, noticing that four ewes that had produced triplets in the morning were now sharing their pens with two lambs each.

'Ah,' said Amy, looking a little nervous, 'we did foster a few on to the ewes, but one or two ended up on Clare and Laura's bottle. They've promised to do all the feeding themselves.'

Out of the corner of my eye, I could just see Laura and Clare at the other end of the sheds. A pair of lambs were guzzling away at the half-litre bottles of powdered milk they'd made up for them. I groaned to myself – the girls had got the better of me again.

It was a relief to have two such competent assistants in the sheep sheds. It gave me more time to deal with the lambing flock at Town Farm, as well as the growing population of lambs and ewes now filling the fields of Borough Farm. For the latest arrivals from the lambing shed, the sharply sloping banks of Steep Sheep still offered the ideal protection against the elements. But with lambing at its peak, it was quickly becoming overcrowded. Every few days now I needed to move sheep further down the farm. As I walked through the fields closest to the farm this morning, there were scores of well 'paired' family groups that could now be driven on.

Turning ewes and lambs out to the pastures at the far end of the farm was always a time-consuming job, so I'd armed myself with three four-legged assistants, Gail, Greg and Swift, and a pair of two-legged helpers, Katherine and Laura. While Gail and Greg prevented the flock from drifting away, Swift helped me cut off well 'paired' ewes and lambs and drive them on to the woodland path. Once three or four groups had been sorted in this way, Katherine took over, driving through the woodland trail that led to the opposite side of the valley.

It was no easy task. While some lambs stick like glue to their mother's side, others wander obstinately in the opposite direction. The dogs too were having a tough time of it. Ewes can be extremely protective of their lambs during the first month after birth, and the sight of the dogs in close proximity was enough to put a lot of them on the offensive. Swift was bearing the brunt of the attacks, but she was also the best of my dogs at dealing with the threats. Her familiar side-step and warning snap of the teeth was enough to send all but the most determined customers retreating.

Chaotic as the job was, we had one important factor going for us. As a rule sheep aren't blessed with a great deal of brains but, for some reason, they have an amazing ability to remember routes to familiar grazing pastures. On the open expanses of hill farms, generations of the same ovine families will live on the same areas of grazing, or 'heft', rarely straying from their chosen area, and always returning there after a gather.

In the same way, the route through the woods is etched on the memory of the Borough Farm ewes. So once the small groups were inside the woods, it took only the minimum of encouragement from Katherine and Laura to get the mothers to trot obligingly along the half-mile of path before emerging on the bank opposite. For Katherine and Laura, positioned at different points along the path, the main challenge was redirecting the odd wayward lamb.

An hour and a half later Steep Sheep was ready for its next intake of ewes and lambs. We walked back towards the farm, hopeful that we might time our arrival back in the sheds to coincide with the latest output from Debbie's oven. No matter what they'd spent the morning doing, the children had an uncanny knack of knowing when their mother was going to arrive with the morning bounty. I found Nick and Clare there as usual today, Clare perched on an upturned bucket, listening intently to titbits about Amy's life at university. To go by the tales Katherine and Amy told, life as a vet student was never dull.

Greg joined us for morning coffee as he often did. He had developed a technique of staring so intently at the cake held in a chosen victim's hand that they were almost shamed into giving him a bite. Today Katherine was the focus of his attention. Once again the tactics were proving successful and she was feeding him a piece of his favourite chocolate cake when there was a disturbance somewhere behind us at the back of the shed.

I turned round to see that some sheep had spotted a gap in the rather tatty steel gates that made up the back wall of their pen, and had made a break for freedom. They had apparently decided that the freshest grass was at the top of the quarry-like bank beyond the sheds and had scuttled there in a chaotic scramble.

'Look back,' I whispered to Greg.

In an instant the cakes were forgotten and he was running out of the shed and up the bank in pursuit. He quickly had the escapees under control. Their plan foiled, the ewes were soon bolting back to their pens. When the last of them had been seen safely back through the broken gate, Greg returned to Katherine and his chocolate cake as if nothing had happened.

Greg responded to the difficult 'look back' command better than any of my dogs. But at times, it seemed to me, his understanding of what was required was almost telepathic. Here I'd only needed to give him the one command and he'd cleared up the mess in moments. Even I was impressed.

'How did he know to do that?' Katherine asked.

'Instinct, experience – and a lot of training, of course,' I said, before giving the girls a brief description of the basic commands and principles of sheepdog training.

'What's the cleverest thing a dog's ever done for you?' Amy asked.

'That's a tricky one,' I said, stumped momentarily. 'Greg's pretty bright, but so was my first dog, Kim.'

I had the girls' undivided attention, and recounted an episode from my early days as a shepherd in Kent. Of all the dogs I'd owned, only Kim had come close to the level of intelligence Greg had displayed this morning. I'd been working late one evening in a huge lambing shed, with a big, wide central alley. The pens on either side of the alley were divided by feeding troughs, designed to be just wide enough to walk through.

As I worked at one end of the shed, I became aware of the noise of wood being gnawed coming from the opposite end of the building. I walked to the other end of the shed and discovered it was coming from Kim, who'd rather bizarrely walked down through one of the feeders. She was standing, nose down, staring at the wooden base of the trough then tearing at the wood with her teeth. 'This is a bit strange,' I thought to myself. But I had such respect for Kim, I knew she must have had a reason for behaving this way.

I climbed into the pen and peered under the trough. As the straw had built up under the feeder, it had left a small cavity. In that cavity was a newborn lamb, completely entombed and invisible to the outside world. The fact that Kim had heard the faint muffled bleat was amazing enough. But her instinct to walk through the feeder and chew through the wood to reach it was truly extraordinary. It showed a thought process far beyond anything that could be trained.

'I was about to turn in for the night. The lamb wouldn't have lasted until morning,' I told the girls.

A faint, furtive hint of pale green was slowly returning to the valley, infusing the morning breeze with a fragrant sweetness. During the middle of the day the buzzards had begun to soar on thermals, high above fields that were once again speckled with sheep. Borough Valley was at last awakening from the long wet winter.

Lambing is never trouble-free, but – thanks to some good help and kind weather – this year seemed to have passed with fewer difficulties than usual. Over half the flock had now given birth, and lamb numbers still seemed pretty good. I estimated that for every one ewe that was rearing a single lamb, there were three raising doubles. If that carried on it would make a successful year, as far as numbers were concerned at least. We were on target for around 1,300 lambs.

Katherine, Amy and myself were working a rota system that ensured we kept an almost round-the-clock vigil. Despite the long, hard slog, the students were learning much. I sensed they were understanding what makes lambing a special time for most shepherds. Being responsible for bringing so much new life into the world is an enjoyable and at the same time humbling experience, even if for the most part you don't have the time to appreciate it.

There were harsher lessons to learn too, however. Not all lambs arrive in the world successfully, and every sheep farm has its share of losses. Perhaps the most frustrating and needless way to lose a lamb is when it is born 'in the sac'. Frequently lambs are born with their nose and mouth covered by the thin membrane that held the embryonic fluid inside the womb. If the ewe is busy fussing over a sibling, or starts licking the new arrival at the back end, the lamb can easily suffocate before it even has the chance to draw its first breath. It is easy to lose a lamb by turning your back for five minutes. But thanks to the vigilance of Katherine, Amy and myself, we had kept the number of lambs lost 'in the sac' to just half a dozen.

Our greatest frustration had been the ewe who occupied pen number 22. She had produced an enormous single lamb two days earlier. At over sixteen pounds he was a magnificent specimen. However, the ewe had shown no maternal instinct whatsoever, and had spent her first day of motherhood either butting her newborn aggressively, or jumping from her pen.

In the meantime, of course, the lamb was getting hungry, and Amy and Katherine had taken it in turns to hold the ewe while the lamb fed. Often, once a lamb has fed from her a few times a ewe will suddenly discover her mothering abilities. But it wasn't to be with this one, and as I'd worked my way through the pens of ewes with lambs the previous evening, I'd found her splendid lamb flattened and dead in the straw of her pen. The ewe had clearly lain on it. Whether it was a deliberate act or not, from the way she now stood there chewing some silage, it was clear she was completely unconcerned.

With several 'spare' lambs still looking for new mothers, I had been loath to let her leave the shed with a full udder of milk, and not feeding a lamb. So earlier today we had moved her through to a 'stock' pen, where her head had been clamped so as not to allow her to smell the lamb feeding from her, let alone attack it. Usually after a day or so in the stocks, a sheep forgets her objection to a lamb and thereafter lets it feed.

Having secured her head, we found a strong, week-old, 'spare' lamb, and placed it into her pen. When I'd checked an hour later all had seemed well – indeed the ewe now seemed to be putting up less resistance to the would-be foster.

It was late in the afternoon that Katherine came to find me, wearing an ashen expression.

'That ewe in pen 22,' she said. 'She's lain on the foster lamb as well. It's dead.'

Katherine had learned much in her time with us, but the realisation that ewes were capable of infanticide was probably the most surprising lesson yet.

'Don't worry,' I told her. 'There's nothing much you can do to stop a ewe that wants to do that.'

The ewe soon got her way. She clearly had no intention of rearing a lamb this year, so a couple of hours later I turned her out to the fields, where she ate voraciously at the grass as if without a care in the world. She was equally oblivious to the red ear tag that she now wore in her left ear, a reminder to me that she was to join the next consignment of sheep for market.

Ernie, on the other hand, was going nowhere, I had decided. Despite all the hours of agonising, weighing up the pros and cons of the situation, I'd found it well nigh impossible to make a decision about his fate. I just didn't have the strength to say goodbye to him. Instead for the past few weeks I'd kept him kennelled for most of the day and exercised him only under close supervision in the evening or early morning. With lambing coming to a close, however, I'd finally settled on a course of action. I couldn't face the thought of losing him. When things calmed down and I had more time I'd step up his training once more. I'd do my best to bring out the natural herding talent that he so clearly had in abundance. I knew he had it in him, but I also knew that he was in the last chance saloon. One more mistake and I wouldn't be able to avoid the issue. It would be the end.

CHAPTER FIFTEEN

Cold Comfort Farming

The twelfth of April, and the first of the swallows had returned to Borough Valley. Ordinarily their arrival back from Africa would be the surest signal that winter was well and truly behind us. But as I glimpsed the first pair flitting around the old outbuildings this morning, I wondered whether, for once, they'd got their dates wrong.

The bitter east wind had returned with a vengeance at Easter, bringing driving rain and threatening to undo all the good work of the last six weeks. Even in the sheds, the last of the season's lambs were hunched tight against their mothers' flanks, desperate for protection from the penetrating cold. However, the welfare of the thousand fragile lives out in the fields had become my greatest concern.

Wrapped up in a balaclava and waterproofs, I set out into the driving rain on the quad bike, quietly dreading what the long morning ahead might hold in store. During these first few weeks of their lives together, the familial instincts are so strong that ewes and lambs are rarely separated. A hungry or lost lamb tends to stand out. With this morning's horizontal rain, however, motherless lambs were seemingly everywhere.

While heavy-coated ewes grazed oblivious to the gale, dozens of lambs had been driven to take shelter under the overhanging hedgerows that grow out of the stone banks. As I arrived in the first field, the noise of the quad bike sent most of them running, calling their mothers as they went. It was the few who remained behind who demanded my attention. One blackfaced Suffolk lamb made no effort to get to its feet as I approached. I picked it up and found its whole demeanour was subdued, almost lifeless. Removing my glove, I put a finger in its mouth. Its tongue was cold to the touch, a sure sign the lamb was chilled to the bone and needed immediate attention. I tucked it inside my jacket, checked the rest of the field for lambs in a similar condition, then headed straight back for the sheds.

Shepherds have always faced the task of reviving hypothermic lambs. One of the older generation of local farmers claimed he once resurrected a lamb he had dug out of a snow drift by putting it in a bucket of warm water and feeding him whisky and egg white. 'He lay there for three days before he started coming round, but he grew like a mushroom after that,' he told me.

Thankfully, today veterinary science has come up with some perhaps more reliable cures. Back at the shed, I took the sickly Suffolk from under my coat, drew up 50 millilitres of a sterile glucose solution and injected it through the wall of its belly into the peritoneum, where the sugar is most easily absorbed. I then placed it in a 'hotbox', a home-made wooden box with underfloor heating provided by a blower fan, where the lamb could warm slowly over the next few hours. Radical as it sounds, this treatment can be highly effective, often transforming comatose patients in a miraculously short space of time. Unable to do any more for now, I left the lamb inside the gently humming box.

I'd lost half an hour and had barely started on the morning's work. I still had several hours to do at Borough Farm, where seven of the eight fields had yet to be checked. After that, the lambs at Town Farm would have to be checked as well. The next few fields told much the same story as the first, with shivering lambs emerging from every corner. None, fortunately, had suffered the conditions so badly that they needed treatment, but they still needed reuniting with their mothers, another time-consuming task.

In a field bordering the woods at the lower end of the farm, however, a different problem soon became obvious. A ewe was dragging her back leg painfully. Her head and ears were hung low, and her flank looked hollow, as if she'd not been grazing. Greg and Swift instinctively jumped off the bike and ran to hold her up for me. The ewe showed little inclination to challenge them.

I was fairly certain of what her problem was likely to be. On turning her over, I saw her udder displaying a large area of bruised-looking purple tissue. When I gently squeezed her teat, it produced a cold, watery, blood-like liquid. She was clearly suffering from one of the most unpleasant sheep ailments – black udder,

It had taken me a while to forgive him. For a long time after the attack on the ewe I had felt let down by him. He was a dog I considered to be of great intelligence. For want of a better phrase, he should have known better. At the same time I bore part of the responsibility myself. It was up to me to make sure he didn't reoffend. I had to make sure he never found himself in a position he didn't understand and make the wrong decision again.

At the same time, however, it was only through experience that he was going to learn and develop into a successful working dog. Today, with the weather improving, I'd taken him out to bring some sheep in from the banks. Part of the flock was grazing among the fresh, pale green bracken where the fields met the woods. A few weeks earlier, it would have been a struggle to stop Ernie taking off of his own volition. Now, although still sorely tempted to follow his instincts, he waited patiently for me to give him his command. On my 'come bye', he set off of a wide arc to the right, the correct line to bring him up gently behind the sheep.

As he bounded through the bracken, a magnificent brown hare bolted from under Ernie's nose. For a moment he slowed, obviously distracted. I gave him a blast of the whistle to remind him where he was supposed to be heading, and he was off again. While Ernie continued in the direction of the sheep, I couldn't resist watching the hare speed away. After a couple of hundred yards, as if sensing he was out of danger, he stopped and sat on his hind legs, sniffing the air.

I lost track of time for a moment. When Ernie reappeared with the ewes nicely under control it occurred to me that he'd taken a bit longer than he should. There could have been any amount of explanations, I thought, already fighting the idea he may have been up to no good again. But when I got closer to Ernie, I was filled with a real sense of dread.

I immediately spotted blood around his mouth. It had also splattered on to the white of his collar. I called him to me, fearing the worst. Already I was bracing myself for the inevitable. If he had bitten again, I knew that this time there could be no reprieve.

'Please don't let him have done it again,' I said to myself, as I bent over to take a look at him. Ernie sat at my feet, complete trust in his eyes as he looked up at me. I couldn't bear to return his gaze.

The sense of relief I felt when I spotted the cause of the blood was enormous. Somehow Ernie had bitten his tongue while running. It was only a tiny nick but, with his tongue flopping loosely from his mouth, it had been more than enough to spread blood liberally around his lips, front legs and collar.

'You had me worried there for a second, Ernest.'

It was late afternoon before I had a chance to get back out to Morte Point. On my way to the sheep I was pleased to see Philip, the National Trust warden, repairing a section of stone wall bordering the Point and the village. A few weeks earlier, a couple of particularly determined ewes had climbed over this weak point and had ended up in one of the village's gardens. I'd borne the brunt of the owners' anger – but it had fallen to Philip, or more accurately the Trust, to fix it.

Although there could have been conflicts of interest between us, our relationship had always been good. Philip appreciated my needs as a farmer, and I understood the Trust's environmental concerns. We discussed any issues that arose and always seemed to come up with a mutually agreeable solution.

'Glad you've finally got around to mending my wall,' I said with an ironic arch of an eyebrow.

'I've got another two and a half thousand acres to look after as well as yours, you know,' Philip smiled.

We chatted briefly, but I was anxious to get on. The odds were against it, but I still hoped the two errant lambs may have reappeared. On locating Red 23, however, I saw she was still grazing alone. The fullness of her udder confirmed that no lamb had fed for the last day or so. To my disappointment, the second ewe too was still unaccompanied.

The rest of the twenty-five ewes had spread out slightly in the past twenty-four hours. I counted them with growing apprehension. To my astonishment, a further three lambs had gone missing overnight. There were now only twenty lambs – a fifth of the group had vanished.

I'd never experienced losses of this level. The evidence was pointing to an obvious conclusion, and the two prime suspects capable of carrying off a lamb. There was an active badger sett on the north side of the Point but, in reality, badgers are opportunistic predators and would have been unlikely to strike so many times in such a short period. Foxes are more cunning and determined, but I still found it hard to believe that one had been capable of taking five lambs from under the noses of their mothers.

Whatever – or whoever – the culprit, I couldn't risk leaving the remaining lambs out on the Point any longer. As I returned to the Land Rover the light was fading fast once more. There was nothing else I could do tonight – I would return with the sheep trailer and collect the lambs and ewes first thing in the morning.

CHAPTER SIXTEEN

Questions and Answers

I was at my desk when a knock on the door interrupted me shortly after breakfast. I knew who it was immediately – one of the year's least welcome visitors, ready to put me through one of the year's least endearing experiences. The farming inspector had made an appointment to visit us a few days before. Typically, he'd arrived twenty minutes early, denying me the vital last few minutes I'd allocated for bringing my chaotically kept records up to date.

The increase in farming inspectors has reached epidemic proportions in recent years. There is now, it seems, a never-ending list of organisations overseeing the welfare of everything from the soil to the stock to the rights of the consumer. Many of them, of course, provide a valuable and necessary service. A few, however, seem to have been designed to come up with a list of rules and regulations that are almost impossible to follow. One or two – including the one concerned today – also send out the most inappropriate inspectors, individuals who, to put it politely, are uncorrupted by a practical knowledge of farming.

I was once visited by an American doctor of something or other who apparently made his living selling organic chillies by mail order. He had seen on his form that I kept Romneys out at Mortehoe and, on seeing the first group of ewes that morning, had declared enthusiastically: 'These Romney sheep are just great, aren't they.' I didn't have the heart to tell him they were Mules.

Today's inspector seemed equally ill informed. To make matters worse, he wasn't as enthusiastic either. He was a nondescript-looking chap, in his early forties, who clearly relished his position of authority. As he stepped into the kitchen, he declined a cup of tea and briskly insisted we press on with the business of the day – an annual inspection of the livestock, the land and the farm records.

In an attempt to win him over, I began by taking him out to Morte Point in the Land Rover. I hurriedly cleared a collection of spray cans, baler twine and coffee cups from the passenger seat before inviting him on board. His rather dour mood wasn't improved as I made a sharp turn right out on to the main road and the Land Rover chose this moment to fling its left-side door wide open. 'Sorry,' I said, hitting the brakes and trying to make light of the moment. 'It's got a nasty habit of doing that.'

He failed to see the humour of the situation, and remained stony-faced throughout the five-minute drive to the Point. Undeterred, I felt sure that the experience of walking some of the finest scenery in England must improve his mood.

'Do you farm yourself?' I inquired, as we made our way through the cemetery gates and the open expanse of the headland revealed itself.

He seemed almost to sneer at the mere suggestion. 'No.'

I tried another tack. The organisation he represented was dedicated to enhancing the environment. I felt rather proud that my flock's grazing of the area's grasslands played its part in creating an environment in which the Point's diverse flora and insect life was thriving. As we walked I pointed out a flourishing expanse of heathers, stonecrop and wild thyme, explaining that it would probably have been swamped by long grasses if it hadn't been for the gentle foraging of my ewes.

Once more, however, my conversational seeds fell on stony ground. 'Hmmmm,' he grunted, as if he'd hardly even heard what I'd said.

After a further ten minutes or so of deafening silence, we stood inspecting some sheep overlooking the rocks towards Bull Point, the next headland to the east. The peace and tranquillity of the coastline was breathtaking – or so I thought. As he officiously ticked a box on his list and headed back towards the Land Rover, his only comment was, 'Cold here, can we go?'

Part of his remit was to inspect all of the land under my management. I drove him round Morte Point for another quarter of an hour before he decided he'd seen quite enough.

'Can we see the rest of the stock now?' he said tersely.

Having failed to impress him with the finest countryside in the land, I moved on to Plan B. I drove him to the fields adjacent to Morte Point, then took Greg and Swift, sending them in either direction to gather the flock that had scattered over the forty

Unfortunately I knew the tranquillity would be short-lived. For the first time since they'd been born, the 250 or so lambs grazing on the far side of the farm needed to be brought into the sheds for the day. With the lambs now into their third month, they needed to have two treatments administered, one for worms, the other for flies. Driving the lambs and their mothers to the handling pens always provided one of the stiffest challenges to the dogs. It was a job to test their stamina, more than anything else. These lambs had, after all, spent all their young lives in just one field.

I'd brought Greg, Swift, Gail and Fern with me to try to minimise the inevitable chaos, but it wasn't long before the lambs fulfilled all my worst expectations. As the sprawling flock neared the gateway that marked the entrance to the path back through the woods, a gang of about fifty lambs started forming at the back of the flock, looking for an opportunity to break back up the hill. As usual, the ewes at the front were reluctant to enter the woods for fear of leaving their lambs behind, and the lambs at the back, separated from their mothers in the confused mass, were determined to run for home.

Sheepdogs dominate their flock through respect. Respect was not, however, a concept these lambs had yet developed so as they ran relentlessly back up the hill from which they'd come, they did so pursued by four increasingly exhausted dogs.

Some dogs are better suited to working with lambs than others. No dog had been better at it than my first, Kim. My second job as an employed shepherd had involved me working with a much bigger flock, nine hundred sheep in all. On a hot June day I set about gathering this enormous flock with just Kim as my helper. It was a flock that hadn't seen a dog for some time so behaved erratically. Kim worked tirelessly, and by the end of the day we'd managed to get most of the sheep in. That evening she was inside in our house, coughing and breathing in a laboured way. She'd worked herself to a point of near exhaustion.

Greg had been a master at dealing with these upstarts in his younger days. But now he was finding the sheer, strength-sapping effort of endlessly chasing after a disrespectful lamb a bit much. Every few minutes he would lie in a patch of long grass, panting heavily, trying to recover his strength. Fortunately, Fern was proving herself a natural at lamb management.

Just as the final group of half a dozen lambs were seemingly heading towards the entrance to the woods, one particularly rebellious gang member fled back up the hill with a speed that would have outpaced even a dog fresh from its morning kennel. On hearing my command, Gail, Greg and Swift looked almost forlornly up the hill. Their tongues were hanging to the ground. Thankfully for all four of us, Fern still had energy to burn. She took off in pursuit and had soon passed the escapee a hundred yards up the hill. The lamb spent a fruitless few moments trying to zigzag its way past Fern, but it wasn't long before Fern had established who was in charge. The lamb was soon sprinting back down the banks in search of its mother, an important life lesson learned.

After twenty minutes or so we'd managed to send the first of the flock successfully on their way along the woodland path. But back on the grassy slopes above, a slowly diminishing gang of lambs were still belligerently refusing to follow. Every few minutes we would, between the five of us, manage to corner two or three of the breakaway lambs and set them on their way, but it was still half an hour before I finally shut the gate on the field and followed down the track accompanied by four shattered dogs. Even then, the good work was almost undone as Gail and Swift, unable to resist the lure of the water any longer, ran past the back of the line of sheep and collapsed into a cool pool, where the path crossed the stream. With the bathing dogs now cutting them off, the back section of the flock turned back up the hill once more in the wrong direction. I gave a few sharp words to the offending dogs, and reluctantly they dragged themselves from their impromptu bath and staggered back into position.

The woodland path offers the most direct and quickest route to bring flocks grazing the fields on the west of the Borough Valley back to the farm for treatment. But it was still an hour after the start of the drive before we finally drove the last of the lambs into the handling pens at the farm.

Administering the treatment they needed only took a couple of hours. Young lambs inevitably pick up stomach worms from the pasture, to which they have no immunity at such a young age. A dose of worm 'drench' given via the mouth prevents a damaging infestation. Blowflies present a much more unpleasant and even more serious threat to the sheep population over the summer, and are the curse of every shepherd on lowland farms. Throughout the summer months these flies lay their eggs on any dirty wool on the sheep. Two or three days later the larvae or maggots hatch and begin eating into the skin of the victim. The sheep's first reaction is to seek sanctuary from the relentless attention of the flies, often by taking themselves deep into clumps of gorse or scrubland. Here the maggots feed voraciously, all the time attracting still more flies to lay still more eggs, with truly appalling results.

Up until the last few years the only preventative measure has been to dip the sheep, completely immersing them in a solution that stays in the wool and keeps the flies away for ten or so weeks. Thankfully, a spray has recently become available which, when applied along the sheep's back, is equally as effective. It's quick and easy to use, and today it took only fifty minutes to treat the entire flock of lambs.

I should have spotted the shifty look on his face the moment Andrew walked into the yard that lunch time.

'Couldn't just sign this for me, could you?' he said, waving a folded piece of paper in my direction.

'What's this, another dodgy reference?'

'Something like that.'

It was only as we sat down with a cup of tea in the kitchen that I caught sight of the paper's letterhead – the insignia of the International Sheepdog Society.

'What's that I've just signed?' I said, snatching it out of his hands.

'Well, I knew you wouldn't enter, so I thought I'd do it for you,' he mumbled.

Andrew had seen Greg and Swift working together and had been adamant that they were good enough for the 'brace' at the annual English National trials, due to take place at Exeter in July. I wasn't so sure. I'd only competed in a modest number of single competitions in Devon and Cornwall with sporadic but unspectacular success. I didn't feel ready for the step up to the biggest and most prestigious competition in the country.

'I'll think about it, Andrew,' I said, slipping the entry form into my jacket and heading out into the yard. 'I've got work to get on with.'

I tucked the letter into the glove compartment of the Land Rover, my mind pretty much made up already that it would remain there for good.

The drive back towards the woods was slightly less chaotic. Sensing that they were returning home, the lambs were more inclined to keep up with their mothers. Even so, by the time the evening came and we had brought a further three flocks in for treatment, the spring had gone from Greg, Swift and Gail's work. Every command, it seemed, was being answered reluctantly. Fern on the other hand seemed less affected by the day's exertions, and still ran ahead as if looking for more work. Stamina varies greatly in dogs, and I was beginning to realise that Fern had been blessed with exceptional reserves of it.

The warmth of a glorious May day evaporated and the sea mist once again crept up the valley towards me as I shut the woodland gate for the last time. A small clump of late-flowering primroses, still bright yellow in the lengthening evening shadows, caught my eye. They marked the spot where, a couple of years earlier, I had buried Kim.

Kim had been a permanent companion for me during my early shepherding career. She'd been responsible for showing me what a good sheepdog was capable of and ignited a lifelong passion. Her loyalty had been exceptional, even by the standards of her breed. When Debbie and I left on our honeymoon, Kim, who had attended the wedding reception, took root in the back of our car and had to be carried out. It was the first time that I had been away without her.

As dusk set in, her four most recent successors ambled wearily in the long grass ahead of me. I spent a moment thinking of the working life we'd shared together, before heading up the hill for home, leaving Kim in the quiet of the evening.

CHAPTER EIGHTEEN

Keep 'em Coming

By the time I arrived at our first shearing job of the season, Geoff and Chris had already set up their equipment. I unloaded my kit from the trailer and set up beside them, suspending the electric motor that powered the shearing clippers on a six-foot stand then changing my wellies for the lightweight shearing moccasins worn for the job. Finally, as if to signal I was ready, I removed my watch, tying it to the stand.

'Why the hell you bother wearing that I'll never know,' Geoff said, casting me a sideways smile between sheep.

'How would I ever know I'm late?' I replied, as I fixed the 'cutter and comb' blades on to the clipper head ready for the first sheep.

Despite the promising start, much of early May had been cold and wet. Not a sheep had been shorn. Now, with a week of warm weather, suddenly every farm in the county wanted their sheep shorn on the same day. Our first venue was a coastal farm, high above Lee where another National Trust tenant grazed four hundred or so ewes. Harold was an older-generation farmer, and was one of my favourite customers. He was in his seventies but still had a faintly mischievous smile and an almost boyish enthusiasm for his flock of sheep. He was always interested in what you were doing.

'Have much luck with your lambing this year?' he asked me as I made my final preparations.

Harold had spent a busy morning, gathering the flock ready to be shorn, ably assisted by 'young Horace', his septuagenarian assistant. Neither man was much more than five feet tall, and both wore flat caps, in Horace's case to cover a completely bald and weather-browned head. The thicker-set of the two men, Horace was there to do the labouring as far as Harold was concerned. Today his main job was as a 'catcher'. I'd been surprised that in Devon the shearers were not expected to 'catch' their own sheep. In this corner of the country, a catcher was on hand to bring the sheep to the shearer's board and thus keep things going. It was every bit as hard a job as shearing, so the fact that someone as senior as Horace should still be catching at his age seemed even more of a surprise.

Within a few minutes Harold and Horace had driven the first pen full of twenty sheep to be shorn. Clippers buzzed and within a couple of minutes the first gleaming white sheep had been returned to their companions.

Shearing is a dirty, greasy, sweaty job. The oily lanolin leaves your whole body feeling sticky and unpleasant. At the end of each day the shower water turns a viscous green at your feet. Even with the double-layered trousers most shearers wear, sore red grease spots still form on your upper legs where you've been constantly holding the sheep. It's also the most physically demanding thing a shepherd has to do. Indeed, I can't think of any task in any walk of life that requires such stamina and strength, with the possible exception of rowing the Atlantic single-handed. I always find the first few days of the season the worst, as I adjust to the constant stretching and bending. I go to bed each night a shattered mass of sore muscles and clicking vertebrae.

The shed in which we sheared was better than most. There was a level concrete floor covered by an old carpet that would make life more comfortable and allowed the clipped wool to be kept clean. Harold knelt on the carpet, rolling the fleeces into tight bundles then packing them into the large, plastic hessian woolsacks.

Horace's task was to make sure each shearer was supplied with the next sheep as soon as the last was finished. It would have been no mean feat for a man half his age, but after twenty minutes the strain was showing on poor 'young Horace'. Beads of sweat were forming under the rim of his flat cap, but he had the quiet determination of a man who'd clearly known a lifetime of hard graft and he carried on regardless, without a word of complaint.

Over the ten years they'd been working together, Chris and Geoff had developed a competitive edge to their partnership. Every shearer is acutely aware of the total number of sheep shown on the tally counter mounted on his stand. Neither liked reaching the end of the day's work to discover the other even one sheep ahead. The truth was, neither Geoff nor I could match Chris's silky-smooth shearing technique. But through a blend of stamina, determination and sheer cussedness, Geoff always managed to keep pace with him. Often he would finish a few sheep ahead.

With Horace clearly struggling to keep up, Geoff and I had begun catching a few sheep for ourselves. Horace saw this and began muttering in his deep, Devonian accent. It was hard to hear what he was on about over the buzz of the clippers and the incessant bleating of the flock.

'What's that, Horace?' I asked.

'Wouldn't have done for Farmer Huxtable when I worked fer 'im,' he said, as he brought the next sheep to my board, clearly implying that he thought we weren't shearing 'cleanly', and were leaving too many 'wool lines' on the shorn ewes.

'Don't worry about that, Horace, you just keep 'em coming,' I grinned at him.

This time he didn't say anything, his face remaining expressionless. He was already planning a very different reply.

Some sheep are easier than others to shear. The variations are enormous not just between breed but according to their condition. The object of the exercise is to remove the entire fleece as close to the skin as possible in one complete piece, while avoiding any nicks. It's also important to do this in the minimum time possible so as to cause the least distress to the sheep. The easiest sheep to shear are those with 'bare bellies' and 'open necks'. The loss of wool in these areas is a sign that the lanolin, or grease, has risen in their wool, so allowing the clippers to slide easily over the skin.

By May most of the North Country Mules I keep at Borough Farm are 'bare bellied', making them a shearer's dream. Harold's flock, however, contained a mixture including Exmoor Horns, a breed that can be the polar opposite, a shearer's nightmare.

A few moments after our little exchange, I saw Horace sizing up a new pen full of sheep. Included among them were a smattering of Exmoor Horns. He wasted no time in catching the first of them and tipping it up on my board. Exmoor Horns can take up to twice as long to shear as a good 'bare belly', partly because every square inch of their body is covered in wool. They also present an extra challenge to the shearer because he has to clip inaccessible little areas of wool around the cheeks, eyes, ears and horns. They are fiddly in the extreme, requiring a technique that I'd never quite mastered, and were always slow work.

This one was typically time-consuming. In the time it took me to clean it of all its wool, Chris and Geoff had finished two sheep each and were well into their third. Any relief I'd felt as I let the ewe go disappeared the instant I looked up and saw Horace passing another, equally uninviting-looking, Exmoor in my direction. It's an unwritten rule among shearers to accept whatever sheep you are given, so I kept my head down and ploughed on.

But by the time Horace had fed me a sixth Horn on the trot I had finally begun to realise what was going on behind his blank expression. Geoff and Chris too had cottoned on and were enjoying the entertainment, chuckling away to themselves whenever they glanced in my direction.

By the time a revitalised Horace had sent me the seventeenth 'Horny' on the trot, all three of us were struggling to work on amid the fits of laughter.

'Come on, David, put some effort into it,' Geoff said, looking deliberately at the tally counter. 'I've done thirty-five while you've been mucking around with those few.'

'I saved all the easy ones for you,' I replied. 'I don't know what you'd do without me.'

As I stood up and straightened my back, the aches and pains in it etched unmistakably on my face, I caught Horace's eye.

'You said to keep 'em coming,' he said, his face showing the tiniest crack of a smile for the first time that morning.

I'd met Chris and Geoff soon after I'd arrived at Borough Farm, while they were shearing for a neighbour. I'd been employed for a day as their catcher. I'd learned to shear when I was nineteen and was keen to join a good working gang in Devon.

'Does the catcher get to shear the last sheep?' I'd asked Chris at the end of my day with them.

Having proved myself competent with a set of clippers, I was asked to help them out the following season. The fact that we got on pretty well helped the long, backbreaking hours pass that bit quicker.

Chris is very quiet and laid-back. He is also punctual – something I'd always aspired to. For Geoff, shearing was in his blood. His father, Bill, had learned the art from the legendary New Zealander Godfrey Bowen. Bowen had invented the modern method of shearing in the 1960s and spent years teaching it around the world. It's a testament to his legacy that seventy per cent of sheep in the world are now shorn his way. Bill had brought Bowen's method – which allowed a shearer to get through two or three hundred sheep a day rather than the hundred possible with the old techniques – back to Devon. He astonished local farmers once by shearing more than two hundred sheep in eight hours in Barnstaple Market.

My decision to join the round may have provided me with a little extra income but it had never threatened to make me wealthy – far from it. When I began shearing in Kent in my teens, the money I earned from a weekend's contract shearing was enough to double my pay packet for the week. Eighteen years on, the rates of pay remain exactly the same, seventy to eighty pence per sheep, depending on the size of the flock. It doesn't really offer much incentive for an established shearer, let alone for a youngster to start the incredibly difficult learning process required to do the job. It wasn't surprising that Geoff, Chris and I were something of an endangered species.

Far more valuable, however, was the way in which working with Chris and Geoff had helped me get to know many of the area's farmers. In most cases it had been a real help in getting me accepted by them, although inevitably some were a little more circumspect about the newcomer in their midst. When I'd first joined Chris and Geoff on the round, it hadn't really occurred to me that, as a man of Kent,

I was considered something of a foreigner to those who had farmed the same fields for generations. 'Hoose hee then?' I heard one of our earliest clients mutter to Geoff, having eyed me suspiciously for half an hour.

On another occasion, we were having tea before the evening's shearing session. The farmer there also ran a bed and breakfast. He was talking about the particularly unusual guests he and his wife had staying at the time, when my name came up. 'They talk funny, like what he do,' the farmer's wife said, pointing a thumb over her shoulder at me. I later discovered their guests were Dutch.

A couple of shearing customers epitomised the best and worst aspects of Devon sheep farming. Geoff, Chris and I always relished a trip to one particular farm at East Down, run by the most hospitable of couples, Brian and Liz. 'You won't need to cut sandwiches tomorrow,' Chris would say with a chuckle each night before we headed there.

The farm, tucked away in a sheltered green valley, was as idyllic as any in the area. The farm's stock too was a picture of good health. A herd of magnificent Charolais cattle grazed in the fields either side of the farm drive, while the sheep were almost twice the size of most. Unfortunately, the sheep's bulk meant they were on the heavy side for shearing, and the speed with which Geoff, Chris and I got through them wasn't aided by the lavish hospitality laid on by Liz. From first to last, we were treated to an endless stream of refreshments. Barely an hour would go by before Liz appeared with a jug of steaming tea accompanied by the latest tray of hot pasties and cake. But the undoubted highlight was the lunch we shared with Liz, Brian and the assorted family members who had turned up to lend a hand.

The table almost groaned under the weight of the food, the centrepiece of which was always a huge roast. I learned the hard way that over-indulgence at Liz's lunch table wasn't always a good idea. On one occasion, shortly after a particularly fine meal, I was presented with one of the farm's largest ewes. As I held the ewe with my legs, her head pressing into my stomach as I began to take off the belly wool, she threw her head back with such force that it caught me unawares. I'd been feeling a little bloated after lunch and was caught off balance. I suddenly found myself being propelled backwards. Before I knew it I was lying face upwards on the floor, as the ewe bolted across the pen.

'I told you you shouldn't have had that last bowl of trifle,' Geoff said with a deadpan expression.

Not all jobs were as pleasant an experience as this, however. One Saturday evening, the last job of the day was at the farm of one of my nearest neighbours. The farm was being grazed temporarily by a flock belonging to a part-time farmer, Charlie. He'd waved me down in the Land Rover one day, asking me to shear them for him.

'It's only 180 – we'll knock them off in a couple of hours one evening,' I'd told Chris and Geoff, almost dismissively.

We turned up at five o'clock on the Saturday afternoon, fully intending to be packed up and gone by eight. But no sooner had we set up our gear in the yard than we realised we might be staying a little later than that. Charlie had told me that he'd got hold of these sheep for next to nothing. It was obvious why. The motley band of geriatric ewes now gathered before us looked like a collection of hat racks. Their shoulders and hip bones were protruding as they trudged lethargically towards the shearing pen. The flock had their lambs with them. It was clear the job of rearing them through the spring had taken its toll.

Their poor condition had had one crucial effect on the wool. There was no hint of any grease in their fleeces, and as we each tackled the first sheep it was soon apparent that the wool had no intention of being separated from its owner. Instead of sliding effortlessly across the skin, the clippers had to be forced with every stroke. When a sheep is shearing well, it's like passing a hot knife through butter. In this case it was like trying to cut frozen butter with a rolling pin.

'Perhaps it was just that one,' I said to Geoff as, after struggling with the first for five minutes, I caught my second ewe from the pen.

Geoff was clearly in an optimistic mood. 'There's a few open necks in there. They'll shear all right,' he said. His positive thinking was admirable – but hopelessly misplaced.

Shearing is never kind on the lower back. But with each of us being forced to spend minutes rather than seconds in the most awkward positions, this was agony.

The pen in front of us hardly seemed to be emptying. By half past six, Chris had reached for his flask and poured a cup of tea. All three of us collapsed on the one full woolsack, arching our spines backwards in an attempt to relieve the pain. Geoff looked at the three tally counters then turned to me and laughed. 'How the hell did you come up with a job like this?'

'How many have we done?' I said.

'I've done thirteen, you've done twelve and Chris eleven.'

By eight o'clock we'd had enough. Between us we'd done seventy sheep in over two hours' shearing.

'There's no way we'll get them finished tonight,' Geoff said, looking at the pathetic tally on his counter. 'I'm not milking in the morning, I'll get here early and make a start.' There aren't many jobs in farming that deserve a medal. But at that precise moment Chris and I would have given Geoff one for services to shearing above and beyond the call of duty.

Charlie had been hovering around, packing up the few fleeces we'd removed into the woolsacks.

'I thought you boys did thirty or forty an hour each?' he said, genuinely surprised at how little progress we'd made.

Nobody responded, but if he'd caught the looks we exchanged between us they'd have needed no translation.

'We shan't be doing these again next year,' I muttered to Geoff as we hauled our shearing motors back to the vans.

CHAPTER NINETEEN

Sennybridge Sales

When Andrew had asked me if I fancied going up to Sennybridge in South Wales, home to one of the country's best-known sheepdog sales, I'd immediately said yes. As it happened, the wet weather had again put paid to the shearing round for a few days and, while neither of us needed another dog, we both enjoyed the day out the sales provided.

The southern part of England and Wales has only a handful of farmers training and selling sheepdogs, with the greatest concentration of them in the hill farms of mid and North Wales. Some of these farmers subsidise their incomes by selling their well-schooled dogs, and with fewer and fewer farmers possessing the knowledge or skill to produce working sheepdogs themselves, markets like Sennybridge now draw large crowds.

The sales field was laid out at the back of the village. A hundred or so cars had spread themselves up the side of the hill overlooking a few acres on the side of a bank where the dogs were sold.

The auctioneer's job was not an easy one. While most customers crowded around his rostrum and signalled their bids with a touch of the cap or a wink of the eye, some had positioned themselves in their cars a few hundred yards away and made their intentions known with a fleeting flash of their headlights or a toot of their car horn.

Fifty or sixty dogs were up for sale as usual. Before bidding begins the dogs are put through their paces with a dozen or so sheep. The variety of dogs on offer, as ever, varied wildly. There were the good, the bad and the plain awful. The buyers' job was to use their expert eyes to spot the telltale signs of flaws in the dogs. At the same time a succession of handlers tried desperately to disguise their dogs' shortcomings – some with more success than others.

As Andrew and I arrived, we saw a two-year-old dog coming into the sales ring. The auctioneer was trying desperately to talk the latest offering up. 'Fine, strong young farm dog,' he assured the gathering.

Everyone there knew that 'farm dog' is a euphemism for an animal that's a bit on the wild side and may never be fully controlled. Nevertheless, someone had made the opening bid of 100 guineas and the price had slowly nudged towards the 250-guinea mark. It was then that the dog displayed a quality the owner hoped no one would see. The collie lunged straight through the dozen sheep, scattering them in all directions. Barking playfully, with his head up and tail raised in the air, he then chased after one of the sheep, driving it towards the fence where the crowd were assembled.

Most of the gathered experts would have spotted the dog's raised head and tail as sure signs of an excitable temperament. This was something no shepherd likes to see in his dogs as it indicates an inability to concentrate properly on the job in hand. In an instant, the bidding stopped, leaving a slightly embarrassed-looking farmer holding the upper bid. The glances flying in his direction were a mixture of amusement and mild pity.

'He's picked a winner,' I said, turning to Andrew.

I've never bought a dog at sales myself. I get more satisfaction from acquiring a pup and building up a relationship with it from infancy. Andrew, however, was studiously scanning the catalogue and the details of the forthcoming lots. The catalogue listed each dog's name, owner and, most importantly, lineage. The ancestry of a sheepdog is hugely important. Rarely is a first-rate working dog produced from a line that doesn't include top quality parents, grandparents and great-grandparents. It never ceases to amaze me how family traits show through. Ernie's youthful obsession with sheep, for instance, was uncannily reminiscent of his mother Swift at the same age.

'That'll be worth watching,' Andrew said in his usual minimalist way, pointing out an entry later in the morning with particularly noteworthy parents.

'I thought you said you were only here to look,' I said.

He just raised his eyebrows and let slip a faint smile.

I knew Andrew was capable of spotting a good dog. It had been four years earlier that I'd asked him to keep an eye out for a well-bred pup here at Sennybridge. I needed a new dog and trusted his judgement.

Pups between the age of six and twelve months are among the most sought-after dogs at sales like this. It's at this age that they begin to show their true potential. At the same time, because they haven't started any serious training yet, it's difficult for their handlers to cover up any weaknesses. Farmers pay up to a thousand pounds for such dogs – a good investment because they're such great raw material with which they can work.

I hadn't wanted to pay that much. But while he was there Andrew had seen a Mr Jones, a farmer from North Wales, selling a slightly older, part-trained pup, aged twelve months. It looked a cracking dog. When he'd got talking to the owner he'd told him he had a litter of pups who had been bred the same way at home, about six weeks old. Following his instinct, Andrew reserved one on my behalf.

About a fortnight later we drove up to Mr Jones's farm in Newtown. He told me I was free to pick any one of the half dozen pups. I had always looked for a pup that was friendly and bold, that came to me rather than shying away. I don't know how I picked her out, but one of them seemed to fit the bill nicely.

Two days later Debbie and I had stood in the yard, trying to think of a name for the new arrival. As we stood there a swallow swooped past us.

'How about Swallow?' I said.

'You can't call a sheepdog Swallow,' Debbie replied. 'How about Swift?'

It wasn't a great surprise today when Andrew and I bumped into Mr Jones as we stood by the refreshment van for a cup of tea late on the morning of the sale. He regularly sold some of the most eye-catching dogs at Sennybridge.

'How'd that pup I sold you turn out then?' he inquired, recognising me.

I was about to tell him what a joy she'd become to work with when Andrew butted in.

'I often wonder how good she'd have been if she'd had a decent handler,' he said, giving me a nudge and a knowing twitch of an eyebrow.

Andrew's dry humour sailed over the poor breeder's head. Instead he looked genuinely aggrieved that one of his potentially finest pups had ended up in such unsound hands.

'Shame, I could have done a lot with a dog like that,' Mr Jones said, shaking his head as he wandered off.

A few minutes later, the dog Andrew had picked out from the catalogue was brought forward into the sales ring. Its owner had a reputation for selling some first-rate dogs, and from the moment it entered the ring it had everyone's undivided attention. The dog was two and a half years old and fully trained. It had a strong, muscular build, and the short hair of a dog equipped with natural stamina. Its prick ears only added to its alert presence. As it began to work, its control of the sheep was perfect. It responded immaculately to its owner's whistled commands. The bidding opened at 800 guineas, and as a hush fell over the field it was soon heading towards 2,000 guineas. Soon the auction had boiled down to a contest between two buyers, one of whom was flashing the headlights of his Land Rover at the auctioneer from the top of the hill a hundred yards away. When the hammer finally fell at 2,600 guineas, a record price for these sales, a ripple of applause ran through the throng. The auctioneer gave a wave to the man at the top of the hill.

'Sold to the flashing lights,' he confirmed.

'Told you it'd be a good 'un,' Andrew said.

'Why didn't you buy it then?' I replied.

It was drawing towards the time when I felt we should be heading back to Devon.

'We'll just see a few more through,' Andrew said, once more running an eye over his catalogue and beckoning me over to a spot he'd seen at the front of the audience.

Next on to the field was a three-year-old white dog. It began working around the sheep, without showing a huge amount of panache. The bidding had begun to falter at just over three hundred guineas, when – to my amazement – I saw Andrew catch the auctioneer's eye with a wave of his catalogue.

'Thank you over there, three twenty,' said the auctioneer.

I'd barely caught my breath before the auctioneer was bringing the bidding to a close.

'Any more?' he said, scanning the thinning ranks.

'Going once, going twice,' he said before cracking his hammer and pointing in Andrew's direction.

'Never thought I'd get her for that,' he said, looking quietly pleased with himself.

We headed over to the rostrum to settle the bill, then set off in search of the owner. Half an hour later, Andrew having collected the dog and been given a run through of its commands by its now former owner, we were back on the road. We'd travelled south for an hour chatting off and on about the day's events, before a look of panic suddenly seized Andrew's face.

'Oh no,' he said.

'What's wrong?' I said.

'I promised Helen I wouldn't buy another dog.'

Summer

CHAPTER TWENTY

The Longest Day

At five-thirty the rising sun was already casting a pale yellow glow over the gorse and heathland of Morte Point. The air was scented with foxgloves, gorse and the fresh new foliage of the season. Only the occasional pip of a meadow pipit flitting from the long grass ahead of me disturbed the stillness.

With almost every other farm in North Devon now filled with gleaming white, fresh-shorn ewes, I'd finally managed to reserve an afternoon to shear the last of my own flock. Geoff and Chris were due to join me after lunch, so for once I was first on duty. I had no complaints; it felt a privilege just to be out amid the serene beauty of the peninsula. With the place to myself at such an early hour, it was one of the genuinely magical times of the year.

The previous night I'd set up a large pen, made out of portable metal hurdles, at the Mortehoe end of the headland. With Greg and Swift, this morning's job was to gather the sheep now spread over the two hundred acres of the Point. I began the well-rehearsed routine by heading for the north side of the promontory. As I reached the coast path I felt the first warmth of the morning sun on my back. Oreweed Cove below me echoed to the noise of two oystercatchers calling to each other.

The first few scattered ewes and lambs we came across were grazing, almost obscured among the deep green bracken, which was by now waist high, having grown at an astonishing rate since late April. I sent Greg and Swift off and the sheep were soon on their way along the path. With the dogs back at my side, we followed.

The temperature was rising fast even at this hour and the dogs were heavily panting already. I let Greg and Swift cool themselves by lying in one of the Point's two natural springs, Morte Wells.

There are only two or three places on Morte Point which cause me real problems when gathering sheep. It was Sod's Law that this morning I found small groups grazing in each of these inaccessible spots. While the dogs had a break, I spotted four ewes and their lambs on the lower edge of Morte Wells 'island', a triangular outcrop connected to the main Point by a narrow causeway. This was somewhere I never relished sending a dog as its natural outrun around the island took it along the western edge which drops vertically to the sea sixty feet below.

I always preferred to send Greg out to this spot. His ability to calculate the danger of situations made him the best candidate. He disappeared off and within a minute was picking his way along the lip of the western cliff. I trusted him implicitly but my heart was in my mouth, as it always was when he made this run. The sheep saw him coming and dived down on to some more gently sloping rocks at the far tip of the promontory. Greg had seen them, however, and I left him to it.

For a moment they all disappeared from view, but within a few seconds the four ewes and their lambs had re-emerged, followed a second or two later by a calm-as-ever Greg. With hardly a command from me, Greg pushed them back towards the causeway. But it was here that his good work was undone. Two of the ewes had been throwing back their heads, obviously looking for a chance to escape. They now spotted an opportunity and shot down over the steep sloping side of the causeway. From twenty feet below, they were soon looking up to see if they'd eluded their captor.

Even from a distance of a hundred yards, I could see the sheep were panting heavily. In this condition there was no way they would walk back upwards for a dog. I headed down the narrow path to the causeway then scrambled down the steep scree bank to where the sheep had taken refuge. After positioning Greg and Swift to block their return to the island, I spent ten minutes persuading the pair back up to the main coast path. By the time I'd set them off in the direction of their companions, I'd lost twenty valuable minutes of my morning.

At least my efforts hadn't gone unnoticed. One of the Point's half dozen grey seals had been watching events inquisitively from the clear, deep waters of the inlet below. High up on the overlooking cliff, a kestrel, guarding a nest of fledglings, voiced his annoyance at my invasion.

By seven o'clock we had cleared the sheep on the northern side of the Point. I was now headed for the highest vantage point, from where I would be able to begin the final drive towards the shearing pen. As we walked around the spine of the Point, I felt the coolness of the sea air being carried on the soft summer breeze. The contrast with the violence of the winter storms of a few months ago could not have been greater. Beneath us the sea now lapped silently, almost benignly, on to the rock.

I was gazing out to sea as I made my way upwards. Perhaps the tranquillity of the scene was to blame, but for whatever reason I failed to notice the coiled form of an enormous adder basking in the warm morning sun. Adders are numerous out here among the rock crevices, but I'd seldom seen one as big as this. I all but stepped on it before I registered its presence. Only a violent lurch to one side enabled me to avoid it.

A few minutes later, from my rocky lookout, I watched Greg and Swift successfully gather the entire flock together to begin the drive towards the pens. It was slow going in the growing heat of the morning. Weighed down by their burdensome winter coats, ewes kept breaking off into the hollows of the gorse bushes in search of shelter from the sun. Each time they did so, one of the dogs would drive them out and back into the main flock. It still took another half an hour to get the sheep back to the pens.

For the final stage of the penning I let Gail and Fern loose. With their youthful energy they proved useful in stopping any ewes escaping at the last. After what seemed like an age, I wearily dragged the pen gate behind the last of the flock.

By now the heat had grown even more intense and both dogs and sheep were panting heavily. A few minutes' rest wouldn't do the ewes any harm, I decided, so I called the dogs out from the shade of a stone wall and headed with them down the steep bank towards a spot from where I knew we could make our way down the cliffs to the sea. As I started to scramble downwards Swift followed easily, but the others were more reluctant. After a lot of persuading, and a helping hand or two, all four dogs had safely joined me in a narrow inlet where the small waves were lapping soothingly on to the rocks. One by one I called my overheated pack to me – then threw them unceremoniously into the sea. Instinctively each of them began swimming back again, but no sooner had they reached the shore than I had tossed them back in again. After a couple of dippings, all but Fern had realised the water wasn't such a bad environment after all. Soon Gail, Swift and Greg were lolling around in the shallows, showing no inclination to move for the rest of the day. The air was cooler and more comfortable here at sea level, so I perched on a rock for a few minutes, washing the refreshing water over my arms and face.

It wasn't long before the faint sound of bleating disturbed the peace of the moment, however. 'Come on, you lot, work,' I called, beginning the long climb back towards the pens high above us.

I'd wormed the lambs and treated them against flies by the time Geoff and Chris arrived a few hours later. Soon after they pulled up in their vans, we were joined by Debbie, who I'd managed to pressgang into an afternoon rolling the fleeces.

The perfection of my morning was soon forgotten as the generator shattered the peace and I dragged the first bulky ewe on to my board. I ran the clippers through her fleece then let her loose back into the shimmering heat of the Point.

Soon the sweat was pouring off the three of us. But with the flock shearing well, we were soon dispatching a steady stream of dazzling white sheep back on to the open headland, and enjoying the sight of a pen that held a rapidly shrinking supply of woolly customers. As the day drew to a close, we were joined by a further set of helping hands. Clare and Laura had been joining in with the shearing and wool rolling at Morte Point since they were toddlers. They seemed to genuinely enjoy having their own role, packing up the fleeces with their mum.

There was a time when wool was so key to the business of sheep farming that this would have been one of the most important jobs of the year. Three hundred years ago our nearest town, Barnstaple, was being built on the profits of the wool trade. On Exmoor there are stories of entire farms being bought with the proceeds of a few years' saved wool clippings. No longer. Today my sheep will produce wool fleeces with an average value of £1.20. It's little wonder farmers aren't prepared to pay shearers a decent wage. When you take into consideration the sheer hard work – not to mention the cost – involved, you can't help coming to the conclusion wool is now nothing more than a nuisance, useful as protection for the sheep in winter but removed in summer for the sake of the animals' welfare more than anything else.

By the time I'd loaded the last of the packed wool into the trailer, Debbie, Clare, Laura, Chris and Geoff had long since left. Even the day's steady stream of walkers had dwindled away. I was alone again.

Looking at my watch, I saw it was nearing ten p.m. I'd been out here for almost seventeen hours, nothing unusual for what was always one of the longest days of my year. It was almost completely dark by the time I clinked the cemetery gates closed.

CHAPTER TWENTY-ONE

The One That Got Away

The sun was still hot in the late afternoon sky as we started back up the Borough Valley and the mile-and-a-half walk back home from Lee Bay. The Sunday walk down to the rocky cove where the valley's waters meet the sea has always been a family favourite. To me the tiny village, with its association with smugglers and wreckers, is a piece of living history. On one side of the breakwater stands an old white millhouse, dating from even earlier than this. Over the centuries its occupants must have seen the very worst that the Atlantic – and human nature – can muster.

For Clare and Laura, however, Lee Bay's greatest attractions are the sea, the rockpools and its supply of skimming stones. The constant erosion of the crumbling slate cliffs has produced a shillet beach filled with the perfect flat stones. While Laura was in the early stages of developing her throwing action, Clare was beginning to really get the hang of it. Nick tried his arm too but was equally content throwing handfuls of pebbles in the general direction of the sea.

As we started our journey uphill, the dark shade of the woodland offered a welcome respite from the heat. The stream at Lee Bay had accumulated water from several springs since it had left our end of the Borough Valley, but as it left the woodland and approached the sea it trickled gently from pool to pool. Nick was long past walking and rode on my shoulders, chatting away.

It's over a mile up the valley to home and no one was in the mood to rush. A game of I-spy had got stuck on the letter V with Laura asking the question. With seemingly every V option rejected, I tried 'variety of woodland vegetation'.

'That's right,' she said, thoughtfully putting us out of our misery. 'Your go now.'

We'd arrived at the point where the path home climbed steeply up and away from the stream and were studying some cloven-hoof tracks in the mud, trying to decide whether they were deer or sheep. There was one set of small tracks and one set of large hooves. Suddenly a rustle in the bramble-covered slopes above made us all look up. We were in time to see a ewe with a lamb disappear into some thick cover.

'Looks like one of ours,' I said, feeling relieved that I'd not made a fool of myself by pronouncing the tracks to be those of deer.

'She's not been shorn,' Clare pointed out.

'How long has she been in the woods?' Laura said, looking concerned.

'Good question – I don't know. But we finished shearing a good month ago.'

Judging by her errant behaviour, the ewe was unlikely to return home of her own accord. Although sheep vary little in their behavioural patterns, I feared this one had decided she was not a flock animal and had reverted to a near wild state.

The woodland covers around three hundred acres of Borough Valley. On the east-facing side, it is almost exclusively deciduous beech, ash and sycamore. The western-facing slopes were felled in the 1940s and 1950s, probably to provide pit props for mining elsewhere in the country. That side of the valley now supports a forest of larch and pine, which has by now grown up to eighty feet in height. The slopes of the valley are steep, and through the summer much of the upper reaches are covered with dense, impenetrable bramble which grows to head height and is entwined with thick bracken. So my chances of finding a ewe in such a large area were a little slim. Thankfully, there were a couple of fences running through which meant that she should be contained in a comparatively small area. I decided to head home with the family now and return in the next day or two with some canine reinforcements.

The early post often has its share of unwelcome surprises, but this Monday morning's delivery had left me feeling a mixture of confusion and quiet dread. Perhaps the mere fact that I'd been sent a programme for the English National sheepdog trials should have aroused my suspicions. But it was when I looked at the very first page that the penny dropped. There in bold letters were the entrants for the first day's competition. Third on the list were the names: Brace: David Kennard – Greg and Swift.

Debbie was already looking a little sheepish.

'What's going on here?' I said. 'The entry form's still sitting in the Land Rover.'

'Well, actually it isn't any more,' Debbie responded. 'Andrew found it and it accidentally fell into a post box.'

'But what about the entry fee?' I said, slowly realising I'd been the victim of forces outside my control.

'Ah,' said Debbie, the guilt now written all over her face. She didn't need to say any more.

I was still trying to work out how I'd been so comprehensively deceived as I made my way out across the yard. It was a few minutes later before I was finally distracted by the fleeting sight of a magpie emerging from the old stone barns. Magpies are a genuine pest as far as most sheep farmers are concerned. Aside from the threat they pose to cast sheep, they are also capable of decimating the bird population. Magpies had severely hit Borough Farm's bird population in the past and my immediate fear was that they were doing so again.

The inside of the eves of the stone barns are favoured nesting sites for the farm's swallows, and we'd had thirteen active nests the last time I'd counted. Most swallow pairs manage to raise two or three broods over the season, with each brood containing four or five young, so there was the prospect of up to two hundred new birds being born this summer.

I approached the stone barns warily. When I'd inspected a nest just inside the door only a day or two earlier, I'd watched parents feeding chicks. Now, as I'd feared, the nest was empty.

Magpies are well known for pillaging nests in the spring, but this was the first time I'd come across a bird rooting out fledglings inside sheds. I'd spotted fourteen magpies in all this summer. If I didn't get rid of this one, there was a danger they'd all pick up his habit, with drastic consequences for the entire swallow colony.

I headed back into the house and to the gun cupboard. I dusted off and loaded my old shotgun, then took up a vantage position overlooking the yard and a spot where I'd frequently seen this magpie. I didn't have to wait long. A loud bang rattled around the yard, and the bird dropped to the concrete in a flurry of black and white feathers. I didn't feel any guilt as I picked it up a moment or two later.

By now, it was about the time the girls' alarm clocks rang for them to get up for school. My faint hope that they'd not heard anything was quickly dashed as Laura appeared at the back door, still in her pyjamas. The sight of me carrying a dead bird in one hand and a gun in the other was self-explanatory. Children raised on farms tend to learn the realities of life and death at an early age. Both girls were passionate about their ever-increasing menagerie of pets, but at the same time lambing in particular had taught them that nature could be cruel.

'Why did you shoot it?' Laura asked quietly.

'It had been eating the baby swallows in the shed over there,' I said.

Laura thought for a moment then, apparently satisfied with the answer, shut the door and went to get ready for school.

The long light evenings of the high summer bring the temptation to work late. It's easy to go on with the endless rounds of worming, foot-bathing and docking, or attacking the summer's growth of thistles and nettles that can spring up from nowhere at this time of the year. The days can be long. Even so, the sound of heavy machinery clattering away in the darkness of the upper fields was unusual.

It had been late in the afternoon that I'd phoned the Met Office for the latest weather report. For the first time that day, a streak of blue had appeared in the west out to sea and I'd been hopeful of fine weather around the corner. Instead I'd been horrified to hear heavy rain forecast for the following morning.

'There's no chance you could bale that field for me tonight, is there?' I'd asked a local contractor, Philip, moments later. 'They're giving rain for the morning now.'

'We'll see what we can do,' he said, unflustered in his quiet Devonian way. 'We've got another four hours' baling here. I'll try and come over after that.'

Baling silage in readiness for winter feeding is a laborious and, to me at least, thoroughly boring part of the farming year. It might have something to do with the fact that my tractor, a fifteen-year-old Ford, had become something of a boneshaker. The prospect of spending dozens of hours bouncing around the fields generally didn't inspire me.

In the two days since it had been mown by a contractor, the silage had dried well and was now ready for baling and loading on to trailers to be wrapped in the airtight black plastic ready for winter. Heavy rain was going to ruin the crop – and set me a serious headache when it came to winter feeding.

It was dark before Philip arrived with his tractor and baling machine. My rusty old Ford was dwarfed by the shining green giant of his John Deere. My 75 horses looked pitiful compared to his 120. Cranking the baler into gear, he was off up the field and a minute later the first of the round bales appeared at the back. The crop coming off was a thick one, and within a few hours 120 bales, four feet high, dotted the field – all of which now needed loading on to a trailer and carting back to the farm.

Fortunately, as well as Philip, I'd been joined by a couple of neighbours, Peter and Adrian. I'd given them an SOS call as well. They'd had even shorter notice than Philip, but had turned up uncomplaining nevertheless.

There was a time when haymaking would have united a village like Mortehoe. Years ago every available pair of hands helped out, as men, women and children moved from farm to farm bringing in the harvest. Remnants of this spirit still survive but, with only a handful of people still involved in farming, the numbers involved have inevitably dwindled. The willingness of my farming neighbours never ceased to surprise me nevertheless.

The previous winter, late at night, I'd been driving home on the tractor when I'd had a puncture in one of the front tyres. With no spare available, I'd been forced to pull into a field and walk for help. Debbie had come out to collect me. Back home I'd phoned Adrian's father Keith, who had a tractor with wheels the same size. I explained what had happened and asked him whether I could borrow a wheel so as I could get my tractor home that night. By the time I pulled up in his yard twenty minutes later Keith had already jacked up his tractor and removed the wheel. He'd dropped everything even though it was a really wet and cold evening. While I don't have a vast array of machinery to offer, I'd always tried my best to help out in the same neighbourly way.

It was pushing midnight by the time the last of the bales were safely wrapped. Even then Philip's night wasn't over.

'You weren't the only one who saw that weather forecast. I've still got another field to bale twenty miles away,' he informed me with a shrug of the shoulders. As the red rear lights of the tractors disappeared into the distance, the summer night sky was darkening and the air freshening. The rain wasn't far off now. I breathed a quiet sigh of relief.

There was a time when a shepherd would have been expected to notice an absentee sheep the moment it disappeared. Not so long ago, a flock of two hundred sheep was all a farmer needed to earn his bread. Each day he would go through the ritual of the morning register, literally counting every head of sheep on his farm. With employed shepherds now expected to look after 1,500 or more ewes on their own, the days when flocks were so closely audited have long since disappeared.

Having spotted the stray ewe two days earlier, however, it was my responsibility to recover her and bring her back to the farm. I headed off with Greg and Swift, as usual, but with Gail as a third dog. Although hunting for two sheep in such a large area of woodland was really a job for one of the more experienced dogs, Gail had recently begun to show signs of maturing. I was keen to give her a chance to learn through experience.

The promised rain had passed through during the morning, leaving the foliage in the wood wet as I brushed through. We followed the stream down as far as the nearest fence, and climbed back up the steep slope of the valley towards the point where the ewe had last been sighted. The wood was quiet and still, only the inevitable buzzing of the woodflies disturbing the peace as I stopped at a vantage spot and scanned the undergrowth hopefully. No sign of a sheep, but my eye was taken by the striking blue of a nuthatch, working its way down the trunk of a large ash tree, picking insects from the bark as it went. I watched it for a moment until it flitted from sight.

I climbed a little higher up the hill to where a small track ran parallel to the line of the fence in the field above. It didn't require the detective skills of a Red Indian tracker to pick up the ewe's trail. I soon spotted a large amount of wool on some overhanging bramble. The path itself was also heavily marked with hoof tracks.

We scrambled over a rocky incline. Immediately Gail dropped her head and adopted an unmistakable pose. She'd seen something. A ewe was resting in a clearing fifty yards in front, her lamb at her side. Her head was down, her ears twitching at the vast swarm of flies around her.

With her superior turn of speed and agility, Swift was the dog best suited to turning the ewe towards home. But before I could send her, the ewe spotted us and bolted flat out with her lamb in the opposite direction. She held her head high, a stance that indicated she had no intention of being caught. I sent Swift off in pursuit, but attempting to pass a determined sheep on a narrow, overgrown path was going to be well-nigh impossible. By now Gail was finding the excitement all too much and, despite my attempts to recall her, she joined her mother. The two dogs were soon disappearing from view.

The chances of turning the ewe were slim. But I left the dogs for half a minute before aborting their mission with a couple of shrill blasts of a recall whistle. I listened for a moment and thought I could hear the sound of returning dogs. But when they failed to materialise I followed it up with a loud 'That'll do.'

Gail and Swift reappeared, panting heavily, their prompt return a sure sign that they'd failed to keep pace with the outlaw ewe and lamb. But the sound of my voice also brought a heavy beating of wings in the branches high above. When I squinted upwards into the upper branches of an ancient beech, I saw the white plumage of the buzzard that had become such a common sight through the past winter. Rejuvenated since the sodden days of November, she swooped through the canopy of leaves above before disappearing from view, the piercing *kiooo, kiooo, kiooo* echoing as she climbed.

Scanning the uppermost branches of the tree, I noticed a cluster of twigs, presumably the buzzard's chosen nesting site for the year. I used the steeply sloping bank to climb until I was almost level with it. Peering through a small gap in the leaves, I could just about make out the form of a well-feathered fledgling, its head tucked down as if cowering from the world. Overhead, its parent signalled its annoyance at the intrusion, calling repeatedly. Borough Valley's small population of buzzards usually produces one successful brood a year. It was rare that I located a nest site, let alone got such a privileged view. I made a mental note of the tree's location, then left the family in peace and returned to my real mission.

We spent the next hour searching fruitlessly for the ewe and her lamb. But they had, it seemed, disappeared. By now convinced that she'd broken back into my fields and would turn up the next time I checked the sheep, I headed for home.

Greg had disappeared a minute or two earlier. I called him repeatedly as we headed back up the track. It was another two minutes before he came back into view, however, and when he did he immediately pulled up. With his ears half cocked, he looked first to me and then in the direction from which he'd just come. His message couldn't have been clearer if he had spoken.

I made the short walk back to where he was, then followed him another couple of hundred yards up the bank. The ewe, with her lamb still at her side, had become entangled in a thick patch of bramble, presumably as a result of her attempts to escape. I took my knife from my pocket and began to cut her free. Greg looked at me and gave half a wag of his tail, as if proud of his find.

'You'rc right, Greg,' I said. 'The rest of them are amateurs compared to you.'

CHAPTER TWENTY-TWO

The Rites of Summer

At seven a.m., the air was already warm and the woods alive with activity. As I picked my way along the path, a thick cloud of woodflies once more swarmed around my head, crawling through my hair and across my face. I was still fighting a losing battle to swipe them away when I felt a pinprick on my arm. I slapped away a horsefly before it could deliver a more painful bite.

The fly season was well and truly upon us, and it wasn't just me who was suffering. In the fields over the past week, I'd treated half a dozen lambs with the telltale sign of 'fly strike'. Frantically twitching their tails and repeatedly 'nabbing' at themselves, they were desperately trying to scratch at the irritation caused by the hatching maggots. The effects of the spray with which they'd been treated in May had worn off. With their wool already thickening again, the ewes too had become susceptible. So the most effective protection now was to take the whole flock back to the farm to treat them, and today I'd got out early, ready for the arrival of the local dipping contractor, Ralph.

With the temperature rising all the time, the morning gather was once more a slow affair. By the time we got the flock back to the farm, all four dogs were already exhausted. Greg, Swift and Fern took the first opportunity to dive into the water tank by the sheep pens. Gail, as was her wont, opted to forgo a soaking and lapped thirstily away instead.

As I drove the last of the sheep into the pens, Ralph appeared from around the corner carrying a hefty canister of sheep dip. Tall, powerfully built, with a barrel chest and a well-trimmed moustache, Ralph was unmistakably an ex-military man. His hair was more silvery now, but in his youth he must have been a formidable figure.

'Morning, David, ready to make a start?' he smiled.

'Ready as I'll ever be.'

I'd spent the previous afternoon clearing out the sheep dip, a six-foot-deep concrete bath into which the flock is plunged. It wasn't the most pleasant job. In the year since it had last been used, the sheep dip had accumulated a couple of feet of slurry that had been washed in with the rainwater from the yard.

Recent legislation means all farmers now face much stricter restrictions on dipping sheep, although the dips appear to have no detrimental effect on sheep. In the past, however, sheep farmers routinely exposed themselves to serious risks while treating their flocks. Up until thirty years ago, the only dipping solution available was a preparation that included, among other things, a small dose of arsenic. In theory, the chemical solution currently used is much safer, although there has been a concerted campaign in the last decade to ban this too. The treatment is based on organophosphates, a chemical whose roots extend back to First World War nerve gas. There is evidence to suggest its use causes nausea, lethargy and damage to the nervous system in some people.

The dangers weren't known twenty years ago. If they had been, I'm sure Judith and a couple of workmates on the farm in Kent wouldn't have thrown me into a dip full of the stuff to mark my eighteenth birthday.

The upshot of all this is that anyone planning to use sheep dip is now legally required to take a government-run training course. The main practical lesson taught here is that you need to wear a large amount of waterproof clothing as protection. Under the mid-morning sun, Ralph and I began struggling into probably the least appropriate outfit you could imagine for a day like today: a pair of waterproofed salopettes, over which we put elbow-high gauntlets and a substantial protective jacket.

Ralph then put a visor over his face and began measuring the viscous, yellow contents of his canister into the four hundred gallons of water now filling the dipping bath. Even outdoors, the sickly smell of the chemicals soon thickened the air. There are a few things that a sheep can remember for twelve months, and the smell of sheep dip is one of them. They immediately started pushing towards the opposite end of the pen with all the force they could muster.

I grabbed the first one and dragged it, back legs first, towards the dip then nudged it sideways with my knee. With the sheep unbalanced, it fell sideways into the dip. Sheep are natural swimmers, and as soon as it entered the bath this one began splashing its way forcefully towards the exit ramp a few feet away. For the treatment to be effective, the sheep must be entirely immersed in the dip and remain in the bath for at least a minute. Ralph, standing overhead with his dipping pole, kept the sheep in until it was adequately soaked.

With 1,700 ewes and lambs to treat in the heat of a scorching hot day, Ralph and I had little energy left for small talk. As the day wore on, the job seemed to become harder rather than easier. The sheep became, if anything, even more stubborn in the broiling heat. The afternoon was a series of endless battles first to force the cussed animals into the holding pen, then to drag them towards the dipper. Underneath the waterproofs, it felt like a sauna. The overwhelming smell of the chemicals was affecting me too. My throat had grown sore from inhaling it and I had the beginning of a headache.

Ralph, on the other hand, was sailing through the day unfazed, even though he had spent the day in closer proximity to the powerful fumes. During the dipping season each year, he might put through 100,000 sheep, yet he showed no signs of it having any detrimental effect whatsoever.

I stopped to cool my face with some water from a hosepipe.

'How you doing, Ralph?' I asked, half hoping he might be showing signs of flagging too.

'Fine,' he said, with a beaming smile.

To Ralph it was clearly all in a day's work. He spent six months of his year bent over the dipping bath, inhaling these dense, eye-watering fumes with no apparent effect. I couldn't help but wonder whether if he cut himself his blood would flow the same viscous yellow as the sheep dip.

'How's Ernie doing, Daddy?'

'Is Ernie being a good boy?'

'Is he listening to anything you say now?'

The phrasing might have been different each time, but Clare and Laura's questioning had become a routine part of everyday life in the past weeks. Since the trouble of March, the girls had been constantly concerned about Ernie's welfare. I'd reassured them he was doing OK, which was the truth. There certainly hadn't been any repeats of that day and I had been pleased with his progress. He had begun to calm down a lot and showed signs that he was prepared to listen to commands, albeit reluctantly. None of this seemed to satisfy the girls' curiosity, however. So today I'd decided on a different response.

'How's Ernie?' asked Laura, half-way through her dinner, as I arrived back at the house.

'He's getting better,' I said. 'You two can come and see how well he's doing with his training tonight if you like.'

Half an hour later we made our way to the paddock behind the house, where I'd got into the ritual of spending half an hour with Ernie each evening. Perhaps it was their mother's influence, but the girls weren't exactly enthusiastic about the intricacies of sheepdog training. As I began explaining what I was trying to achieve, they both glazed over a bit.

They livened up a little when I set Ernie off. They watched intently as he made a good outrun to the sheep, then brought them back to me at a gentle pace. To me it was a particularly good piece of work mainly because he was listening to me a good deal of the time. Far too often Ernie would drift off into a world of his own, where he became obsessively focused on the sheep in front of him. But not tonight.

'Well, do you think he's any good?' I asked the girls, after a few minutes.

'Mmmm,' they replied in a duet.

'Don't you think it's good that I can even leave him off the lead when he's not working now?'

'Yes, it's very good, Daddy,' said Clare in a resigned tone, clearly worried Ernie's training would go on all night unless she gave me the thumbs up.

'So does that mean you're both going to stop worrying about him so much?' I said, sitting down next to them.

'Suppose so,' said Laura with a smile.

It was a glorious evening. Two buzzards soared on the thermals over the valley. One was the bird with white plumage, now accompanied by her fledgling. The two had been ever present for the past few days, their incessant high-pitched calling resonating in the quiet evening air. Tonight, however, they had company – a pair of crows that had taken umbrage at their presence and begun swooping at the buzzards in an attempt to drive them away. It was, we decided, a futile battle. The buzzards, though considerably less agile than the crows, were far larger, and paid little attention to the aerobatic attacks coming from the black marauders.

As the mismatch continued, I wondered how many generations of children had watched this scene. I was lost in my thoughts when the female buzzard turned towards us, emitting a particularly loud *kiooo* as she did so. The sound wasn't unlike the whistle I'd been trying to get Ernie to respond to as his left-hand command. In an instant Ernie was running back down the paddock, half-way through an impressive left-hand arc to where the sheep were once more grazing.

The girls erupted in a fit of giggles.

'I think you've still got a bit more training to do with him,' laughed Laura.

'Can you two give me a hand in a couple of hours?' I said to Clare and Laura as I headed out the door after breakfast.

'Why, what's happening?' Clare asked, her bleary-eyed state lifting slowly.

'I'm weaning, so we're going to have the whole flock in the yard.'

The two of them were well used to helping out on my busiest days.

'OK, Dad, we'll be ready,' Clare said, exchanging a look with her sister.

It's an old shepherding saying that 'July milk does lambs no good'. By this we mean that what little milk the ewes have left three months after lambing is of virtually no nutritional benefit. With the pastures green and lush, now is the time for the lambs to switch to a grass-only diet. The main problem, however, is that if the flock remains intact, then the ewes will continue to monopolise the pastures, growing fat while the lambs lose condition and fail to grow.

If there's a bond between the lambs and ewes still, it's more to do with habit than any physical need. That's not to say there aren't still strong maternal instincts at work, but their ties are now more to do with their being used to each other's company.

There's another good reason why this is the ideal time to separate the lambs from their mothers. It was now a week since Ralph and I had finished dipping. Lambs can't be sold for at least a month after they've been treated. The stress of being removed from their mothers will mean the lambs will lose condition for a while, so they may as well go through it during this time.

Weaning is always a noisy, chaotic job, so it's one to get over and done in a morning if possible. I spent the first couple of hours gathering the entire flock in the field nearest to the handling pens. Collected together in one spot, the flock made quite an impressive sight.

I then began passing the first of the flock through the race in the handling pens. As they passed through one by one, I switched the gates so as to guide the lambs into one pen and the ewes into the other. It's a mesmerising job that requires more concentration than you'd imagine. With the speed that the animals run through, it's easy to get it wrong. And once one goes the wrong way you tend to start switching them all the wrong way.

It took nearly two hours to persuade the whole of the flock through, by which time the noise of the orphaned lambs and their dispossessed mothers had become almost deafening.

If there's one basic principle with weaning it is to get the ewes and lambs as far apart as possible. I took the ewes first. They are usually the easiest to drive. For some unknown reason, ewes always seem to imagine they will find their offspring back in the fields where they last saw them. I took them through the woods into the fields at the furthest end of the farm.

The lambs would be a different matter. I was going to place them in a field at the opposite end of the farm, but first I needed to negotiate the fifty yards from the pens to the fields – in theory a simple enough task.

Clare and Laura had appeared in the yard ready for their instructions. I positioned them so they could block the only possible escape route, up the farm drive. Arms spread, makeshift crooks in their hands, Clare and Laura stood waving and shouting as I drove the nine hundred lambs out of the pens. I'd mustered as much dog power as I could manage. Greg, Swift, Fern and Gail were stretched to the limit covering every move.

Thus far in their lives the lambs had not had to think for themselves. Now, deprived of a leader, they milled round in a swirling mass, hopelessly unable to decide which direction to take. Eventually, however, through sheer force of numbers two or three broke free. Sheep have no respect for humans who pretend to be sheepdogs and, despite the best efforts of Clare and Laura, they were soon streaming past them and up the drive. Now the herding instinct took over. With half a dozen lambs gone, the other 894 headed off in pursuit, and by the time I'd driven the last from the pen there was already a stream two hundred yards long heading in the wrong direction.

'Dad, they've gone up the drive, we couldn't stop them,' Clare shouted, a look of genuine mortification on her face.

'Come on, let's head them off before they get to Woolacombe,' I said.

I whistled frantically to the dogs. In fairness to them, they did their best to bring a little order back to proceedings. But by the time I got to the turn in the drive leading to my neighbour's house, around six hundred of the lambs had already filled her front garden. Bemused, leaderless and now hemmed in by the dogs, they had no intention of leaving. Throughout I kept my fingers crossed that the neighbour wouldn't open her kitchen door at this precise point. Not only were her pansies suffering horrendous damage, there was a very real risk that the lambs might now see the kitchen as a new escape route.

Quickly repositioning the children to stop the lambs bolting even further up the lane, I began assisting the dogs, slowly turning the reluctant lambs in the direction from which they came. It took twenty minutes to drive the flock towards the correct field. With Gail, Fern, Greg and Swift I made a final push to get the swirling mass through the last gate. The finer points of sheepdog craft had long been forgotten by this time.

That night, a light summer's breeze carried the sound of nine hundred despondent lambs and six hundred bellowing ewes through the open bedroom windows. It generally takes the flocks a day or two to get over the separation – and so it proved again this year.

The maternal instinct of some ewes is not to be underestimated. The following morning I was greeted by the sight of fifteen ewes who had managed to cross four banks, eight fences and a quarter of a mile of woodland to be reunited with their lambs in the top fields. I drove them back to where they belonged, taking half a dozen stakes to mend the dismantled fences en route.

But by mid afternoon there was another bleat in the yard. One of the same ewes – a strong-looking, dark-faced Mule with a blue plastic tag in her ear identifying her as a four-year-old and in her prime – was back, having repeated her epic journey. This time I loaded her into a trailer and returned her once more. Again I had a fence or two to repair on the way.

By the following morning the bleating had subsided a little. In the fields the lambs were starting to settle. With Greg and Swift, I was walking along the uppermost, south-eastern boundary of the farm, when I saw – to my amazement – that the same dark-faced ewe had found her way back to her lambs yet again. She'd identified her own two offspring among the nine hundred others and was grazing happily with them.

I'd seen some determined ewes over the years, but this was exceptional. There was something rather touching about her determination. I admitted defeat and left her to it.

CHAPTER TWENTY-THREE

The National

With the sheep dipped and the lambs weaned, it was a good time to take a break. Debbie and the children had been keen to visit the Kent contingent of the family, so we all headed for a week with my in-laws. When we'd moved to Devon, Debbie had left behind two brothers and two sisters, so for her in particular it was a chance to catch up on the family gossip. For the children it was an opportunity to play with their cousins – though for Nick, while he enjoyed being doted over by his extended family, the real attraction of the visit was the shed of enormous machinery on the farm where his Uncle Bob worked. Prising him out of the seat of Bob's combine harvester was well-nigh impossible sometimes.

For me too there were familiar faces I was always happy to see again. With Clare out riding with her cousin Heather, and Debbie visiting her sister Judith with Laura and Nick, I sneaked off a couple of miles down the road to the farm of a shepherding friend, Ron. Ron was becoming something of a rarity in Kent. When I started work there thirteen years earlier there were dozens of employed shepherds on the county's farms. Now he was one of only a handful left. A Scot exiled south of the border for more than twenty years, Ron had appeared rather a daunting figure when I'd first met him. But as neighbouring shepherds we'd formed a friendship.

He was on his way out when I pulled up in his yard.

'I've got to bring some ewes back for dipping tomorrow. Jump in if you like,' he said.

His was a very different farm to mine. His modern shed and tidy yard rather put my disorderly operation to shame. Large flat fields with straight-lined fences stretched for miles in every direction. His flock was made entirely of pure Romneys. The quality of the grazing there had produced big-framed sheep that would have dwarfed my own.

It had been a long time since that part of the country had seen any rain. The fields were parched and browning and by mid-afternoon the sheep were gathered in small groups looking for shade under the few scattered oak and ash.

Ron took one of his dogs from the Land Rover and sent him off to gather the sheep. In the afternoon sun they came up reluctantly, twitching their heads at the cloud of flies swarming around them. One ewe hung to the back panting heavily.

'I see your Kentish sheep still don't like the heat,' I said with a knowing smile.

The smile that crept on to the corner of his face told me he knew exactly what I was thinking.

'Just as well we're not shearing today,' he said knowingly.

The memory of the unfortunate events of another blistering summer's day several years earlier often cropped up when Ron and I got together. We'd been shearing together at another sheep farm in the area. Toiling in the sauna-like conditions, it had been a job just to keep going. The sheep too had been feeling the intense heat. A lot of them had arrived in the pen with their heads down, breathing hard. I'd been half-way through a particularly fat ewe when I'd noticed that she was becoming limp and lacking in the usual fight. A glance at her face revealed she was obviously unconscious.

Heart attacks, although rare, can occur during shearing. The strain on a sheep can be great, particularly in very hot weather. As the ewe lay motionless on the board, I felt sure this was what had happened here. I switched off the machine and called to the farmer, 'I've got a heart attack here.'

Ron was rapidly finishing shearing his sheep.

'I'll give her some heart massage,' he said, dashing over a few seconds later.

Before I knew it, Ron was on his knees by the side of the stricken ewe. Massage wouldn't have been the word I'd have chosen for the procedure he administered, as he began frantically pummelling the sheep's chest with the palms of his two hands. As Ron worked on – shouting the occasional 'Come on, girl' at the still lifeless sheep – the farmer appeared with a bucket of water, which he then proceeded to pour into the sheep's ears, without an explanation.

With all this activity, none of us had noticed that our little drama had acquired an audience. The farmer had neglected to tell us that a party of six-year-olds from the village's primary school had been invited to watch the afternoon's shearing. They'd chosen this moment to arrive in the shearing shed. Quite what these impressionable youngsters thought of a mad Scotsman bouncing up and down on a half-shorn dead sheep while someone else poured water into the wretched animal's ears, I dread to think. Their accounts of their day out on a sheep farm must have made for colourful reading back in the classroom.

'Just drag her round the corner, out of sight,' the farmer said, on spotting his wide-eyed visitors. The sheep was still utterly lifeless. She'd probably been dead even before I'd noticed she was unconscious.

Like me, Ron was an enthusiastic sheepdog triallist. Unlike me, however, he'd had regular success on the trials field, as well as in the breeding world. I'd seen his name down on the programme for the English National where he was running in the singles class with his latest top dog, Spot. As the heat of the day faded, I watched as he put Spot through his paces in a field at the back of the house.

Ron had a distinctive way of issuing commands to his dogs. Rather than using a plastic or tin whistle, like me, he relied on a more natural instrument. By simply blowing through his teeth he could muster an amazing, ear-splitting whistle five times the volume of anything I could produce.

'That's amazing, Ron – not only has your dog stopped, every other dog between here and Canterbury's cowering under their bed as well,' I joked as he sent Spot up the field and blew the first stop command.

I'd seen Ron run several dogs in trials over the years. His strain had always had a directness and strength about them, which made them stand out as genuine working dogs as well as successful competitors. Spot was true to the type. Three hundred yards away he gathered the small flock of sheep with a measured precision that was truly impressive to behold.

'He seems to be on good form at the moment,' I said.

'Aye, he's had a few placings lately,' he replied. 'I see you put yourself down for the brace at the National,' he added after a pause.

'I didn't exactly put myself down, Ron,' I said. 'But yes, we're running. I'm still not sure whether it's a good idea or not.'

Back in Devon, Andrew had taken on the role of manager in the past few weeks, setting up practice courses for me in the evening at his farm. The unfamiliar surroundings would help Greg and Swift deal with the strange environment they'd face at the English National, or so we hoped. There was no doubt in my mind that these two dogs had a very special partnership. Greg's brain seemed to compensate for Swift's weaknesses, while Swift's turn of speed and controllability made her a perfect foil for her more free-thinking partner. It was a natural partnership between two dogs that obviously enjoyed each other's company. But even though I was quietly confident they could perform at the highest level, I couldn't rid myself of the knot in my stomach whenever I thought about what lay ahead. Inwardly I was petrified.

Ron could obviously see my apprehension. He also sensed that I'd be more than happy to accept any pearls of wisdom he might want to share.

'Do your two work well together?' he asked.

'They keep to their own sides well, which helps,' I replied. 'And they both work at about the same pace.'

I admitted my biggest concern was Greg's lack of ability in penning the sheep at the end of the run. Rather than steering the sheep into the pen, he had a habit of letting them slip around the back, something that cost a lot of deducted points.

Ron knew how much more difficult the brace was than the singles. 'If you get that far you'll be doing well anyway,' he smiled.

As far as Ron was concerned, the key to a successful run was to concentrate and watch the sheep rather than the dogs. 'If the dogs are taking the commands, you've only got the sheep to worry about,' he said. 'That and making sure you hit all the gates. Take your time and you'll be all right.'

'I'm glad it's that easy,' I replied, still feeling less than convinced.

The grounds of Powderham Castle in Exeter are an impressive sight at the best of times, but today they looked like something straight out of an English Heritage tourist brochure. Beneath a sky so clear and cobalt blue you felt like simply staring at it, the manicured grounds and well-tended gardens were a riot of colour. It was a shame my nerves were too shredded for me to appreciate the splendour of it all.

A sheepdog trial as important as the English Nationals hadn't been held in the South West for some years, and the hard work put in by the local organisers was obvious. A half a mile or so from the castle itself, a large parkland course had been fenced off for the three-day competition. Along part of its perimeter, the off-white canvas of marquees gleamed in the sun. Rows of brightly coloured stalls, trade stands and catering vans dotted the surrounding landscape. The strange, disembodied sound of the announcer's voice, echoing out of the tannoy, was already reverberating to the four corners of the park.

As I arrived to register, it was another voice that was concerning me. It was a tiny one somewhere in my head that was still telling me to turn around and head home, that I didn't belong here. In the main, however, it was being drowned out by another, more confident voice, telling me, 'Come on, you can do it.'

Compared to those competitors who had travelled hundreds of miles to be there, I was fortunate that I was able to give the dogs a run at home at seven o'clock that morning. It had given Greg and Swift the chance to shed some surplus energy. It had helped me overcome my butterflies as well.

The English National is held over three days, from Thursday to Saturday. While 150 dogs compete in the singles, the more specialist brace competition is restricted to nine pairings, three of which are run at lunch time each day. A win at the National is prestigious enough on its own, but for the leading competitors there is the added incentive of qualifying to represent their country. The top fifteen in the singles and top two in the brace are given the honour of representing England at the annual International final, to be held this year at Aberystwyth in September.

As the competition got under way on Thursday morning, the idea of me representing England seemed about as likely as me winning the Monaco Grand Prix – in my Land Rover. It couldn't have been further from my mind.

The draw had put me in the first group on the first day. I would be going third. At least I wouldn't have to go first, I told myself.

Brace competition in England has been dominated for the past twenty years by one family, the Longtons from Lancashire. Thomas Longton was the current champion with nine wins at the English National, a record he shared with his father Tot. If he won again this year, he would set an extraordinary new record of ten titles. As well as Thomas, his cousin Timothy, another hugely successful triallist, was entered to run at Powderham Castle.

The chance to see the very best dogs in the country competing – not to mention the fabulous weather – had drawn a crowd in excess of a thousand. First to go in the brace was one of the most experienced handlers in the trialling world. With Debbie, Andrew and the children, I watched him stride to the post, hoping to pick up a few pointers for my own run.

The brace is a real challenge for both the shepherd and the dogs. The handler must send both dogs to collect a 'packet' of ten sheep. One dog runs to the left, the other to the right. The two dogs should arrive at the sheep at the same time, then drive the animals around the course and its obstacles, at all times working in tandem, and keeping strictly to their respective sides. Having shed the sheep into their own group of five, each dog must then put its group into a small pen. For the first dog the challenge is even greater as its pen has no gate to shut the sheep in. The dog must remain to guard the sheep while the second dog pens its sheep. Unlike the singles, the brace is divided into seven sections with a total of 140 points available. With two judges on hand today, that figure was doubled to 280 points.

It was soon clear that the first competitor was going to lose a fair proportion of those points rather quickly. His two dogs began their outruns in textbook fashion. But the sheep took one look at them arriving to take control and bolted flat out down the course. It wasn't the handler's or his dogs' fault. They'd just drawn a 'bad packet' of sheep. With the wisdom of his experience, he realised there was no chance and simply walked off the course, his dignity intact.

Any optimism that I had been previously harbouring went out the window. If the experts can't do it, what chance have I got? nagged the negative voice in my head. The second competitor fared a little better – at least the sheep didn't bolt away from his dogs. But it was clear this was a tough trial. Before I had too much time to dwell on it, I heard my name being announced over the tannoy.

I rubbed the dogs' ears gently, hoping it would disguise the tension I was feeling.

'OK, you two, this is it,' I said to Greg and Swift.

Walking to the post at a sheepdog trial is an experience unlike anything in any other sport because there are so many variables on which your success depends. Of course there's the ability and training of your dogs, and how they will react to the particular challenges of the course. But, as the first competitor had just reminded us all, there was also the sheer unpredictability of sheep. Sheep undoubtedly react in different ways to different dogs. How was I to know how the packet I was about to work with would react to Greg and Swift? Then there are the variations within the packet. Are they a group that will keep together or is there one in the bunch that will tirelessly attempt to break from the rest? As I stood at the post, however, these questions were overshadowed by the one burning issue on my mind. What if the handler was so nervous and dry-mouthed that he was unable to get an audible squeak from his whistle?

I knew that to convey my nerves to the dogs would be fatal, so I tried desperately to focus on the job in hand. They stood at my feet, Greg to my left, Swift to my right. I gave my usual quiet whistle to Swift and she was gone, striding purposefully towards the top of the course. Greg stayed motionless until, a few seconds later, he was gone, responding immediately to his command 'Away, Greg.'

As if by magic the two arrived faultlessly at the far end of the field, crossing over exactly behind the sheep, before lifting them gently in my direction. A dream start. The first obstacle to be negotiated was the 'fetch' gates, two hurdles, seven yards apart, directly in a line between the post and the starting position of the sheep, through which the sheep must pass. To the watching judges it was just as important that the sheep keep to a straight line as it was that they pass through the gates themselves.

One of the most difficult parts of brace handling is keeping the dogs to their individual commands. But things were going rather well, with Greg and Swift responding excellently to their own instructions and the sheep keeping an almost perfect line to and through the fetch gates. My confidence began to grow.

The next stage was to return the sheep to me, turning them behind the post at which I stood and then away from me through the first set of 'drive' gates. Again things were looking pretty good. The sheep were a little off line as they turned around me, but they were under good control and Greg and Swift were demonstrating the sort of teamwork and understanding of which I knew they were capable. I was feeling a little easier at this point. There should be no humiliation now, I thought to myself. I'd proved that they were capable of a level of work which justified our entry, at least.

The first set of drive gates were negotiated without any problem. A whistle to Greg, and he flanked to the left to turn the sheep to the right. Swift took her stop whistle and in a second the sheep were running left to right, across the course on good line for the third set of gates.

This next section, the 'cross drive', was the one I found the most difficult. The gates to which we were aiming were 150 yards away, and it was very hard to judge which line the sheep should be taking. On top of this I usually found the dog furthest from me, Greg on this occasion, tended to try to get too far ahead, and in doing so started to return the sheep to me.

As we approached the gates, I wasn't confident that we were 'on line', but I thought we were pretty close. We only had a few yards to go when I realised that the sheep were too far in my direction. I gave Swift the whistle for a half right flank. In the heat of the moment perhaps my whistle wasn't clear, or perhaps it wasn't loud enough. Whatever the cause, by the time I had repeated the command in a more forceful manner it was too late – the sheep passed the wrong side of the gates.

The run had been going well, and up until this point a hush had descended on the audience behind. Now for the first time there was a brief murmuring. Points had been lost, but the situation was far from disastrous. Now it was back to the shedding ring, a forty-yard ring identified by heaps of sawdust on the grass. Here, using only one dog, I needed to divide the ten sheep into two groups of five.

Swift was my shedding dog and, to my delight, no sooner had the sheep entered the ring than a gap appeared in just the right place. I called Swift in, she came to me and held the two groups apart. I quickly counted one group, just to make sure that I hadn't made what would have been a rather embarrassing mistake. I needn't have worried – I had the perfect 'shed'.

Suddenly I felt a new type of pressure building within me. I'd far exceeded my expectations. Now I realised that if I could pen both groups of sheep successfully I'd be in with a chance of the unthinkable – winning!

The first pen was Swift's. Greg had taken his sheep a little way away, and lay down blocking their return. Swift and I took the other group to the first pen. The sheep made a couple of half-hearted attempts to break away, but Swift was equal to them. Slowly we backed them in through the narrow entrance. The sheep looked relaxed. I called Swift to the centre of the pen entrance. 'Good girl, Swiffie, you stop there,' I said, a degree of pleading in my voice as I stroked her head. Her job now was to keep the sheep in the pen while Greg and I penned the second group fifty yards away.

As I gave Greg a command to fetch his sheep, I knew I was now relying on a large measure of luck. Greg is a wonderfully clever sheepdog. I could talk about his attributes for hours. He was a dog on whom I depended day in, day out. He was, however, hopeless at penning sheep.

I can't remember if I said a little prayer at this point, or whether I just pleaded with him under my breath. Whatever method I used, it seemed to be working. As the sheep approached the pen it looked as if they were going to oblige and trot straight in. However, Greg had left a rather large gap on his side of the pen. One of the ewes spotted it and before I knew it all five ewes had headed through the gap and around the back of the pen. In true Greg style, rather than heading them off, he went after them to bring them back. Half the points for the pen were now gone.

We began a second attempt, and straightaway the direction in which they were coming looked more promising. Greg inched forward as I gently tapped my crook on the ground with my right hand to deter any breakaway. The first ewe looked into the pen, and – resigned to having lost the battle – walked in, followed by her four companions. As I slammed the gate shut, the feeling of relief was completely overwhelming.

I looked up to call Swift away from her station but, to my consternation, she had left her post to get a closer look at Greg's sheep. Fortunately her sheep had stayed in the pen, which meant that – as far as the judges were concerned – her wandering off was irrelevant. We had successfully completed the course.

I walked back off, glowing with pride. Andrew and Debbie had lived every second, as had Ron, who had joined them during my run. Debbie gave me a peck on the cheek, Ron and Andrew offered their quiet congratulations. The children cuddled the dogs. As I walked back to the car to give the dogs some water, Thomas Longton walked over and offered his congratulations too.

'How long have you been running brace?' he asked.

'That's my first run,' I said.

'You've got some good dogs there, they've picked it up fast,' Thomas said.

It took the judges an hour or so to put up the day's brace scores. As they did so, I saw that I was leading with a score of 228 points, a respectable total.

I returned to Powderham Castle on the Friday and Saturday and have to confess I watched the rest of the brace competition with a lamentable lack of sportsmanship. My points from the first day had been displayed and I was some way in the lead. All I needed now was for the next six runs to come to grief! I tried hard to wish no ill on the following competitors, but I did feel an undeniable relief when a run hit a problem.

On the third day, however, Thomas Longton himself went to the post. With the assurance of a snooker player making a maximum break, he kept the dogs under perfect control from the moment they left his side. As he slammed the gate shut on the second pen, I was sure he had once again won. Like all top sportsmen, he had put in a great performance when he most needed to.

As I'd suspected, when his points were announced he had taken pole position from me by six points. He had 234 to my 228. He had his tenth title, a quite amazing achievement.

A small part of me was disappointed, naturally. But my overwhelming feeling was one of pride. I'd achieved far more than I had ever dreamed, second in the English National finals. I was delighted. It took a while for the smile on my face to slip, but it did eventually.

It was as I stepped up to receive my prize, a bag of dog food, that I realised that I would now face a far more daunting task; representing England in September's International competition. I could feel the butterflies building again immediately.

CHAPTER TWENTY-FOUR

Disturbing the Peace

At the height of the summer, North Devon attracts visitors in their thousands. For six weeks the beaches of Woolacombe and the paths of Morte Point, in particular, are crowded with people drawn to the area's outstanding natural beauty. For most of the year I feel an almost integral part of the landscape, but during these peak weeks of the summer the atmosphere changes and I become more of an object of curiosity. For a child from the inner city the sight of a scruffy, crook-wielding shepherd with dogs at heel must seem like something from a bygone era. It's little wonder they sometimes stop and simply stare. So at this time of the year I tend to get my regular check of the Morte Point flock over early. I try to be back having breakfast before the first walkers emerge.

Fortunately, August is generally my quietest month of the year. So I'm quite happy immersing myself in the routines of the late summer: worming, foot-bathing and hoof-trimming ewes and lambs in the low shed. It can be an uncomfortable job. The shed becomes hot, and the smell of dung and ammonia makes the work sticky and generally unpleasant. But with the rush of shearing, dipping and weaning over, the pace is more leisurely, at least.

I couldn't hope to hide myself from the world completely, however. There are always plenty of people intent on disturbing my peace. Back early from Morte Point one morning, I was greeted in the yard by a fresh-faced young man climbing out of his rather road-weary Ford. He really didn't need to state the obvious, but he introduced himself as a sales rep for an agricultural company nevertheless.

There are reps whose products I know and trust and with whom I do regular business. Then there are those who try to foist the most unlikely remedies on farmers. I had my suspicions about which of the two categories this young man fitted, and at any other time of the year would probably have given him short shrift. But with things generally quiet on the farm, I gave him the benefit of the doubt. His cause was helped by the fact that he claimed to have a product that might help me deal with the two problems I was currently wrestling with: scald and mineral deficiency.

Scald is caused partly by the coarse, long-stemmed summer's grass. The grass rubs through the foot as the lamb walks, creating a raw area that can become infected with bacteria that live on the pasture. As with many sheep diseases, it passes easily from sheep to sheep so it needs to be nipped in the bud. I'd spotted a few lambs walking tenderly in recent days and, on inspecting them, it was obvious there were small sore areas between the hoof. I'd begun treating them with a walk-through footbath containing a zinc sulphate solution.

A deficiency in two key minerals, selenium and cobalt, is less easy to deal with. It's a perpetual problem for sheep on this coastal stretch of North Devon because, as a general rule, ground close to the sea is of poor grazing quality and lacking in these essential minerals. A lack of cobalt is easy to diagnose. A deficient lamb will become dull and listless, and eventually stop eating. Its ears will peel and look scabby. A lack of selenium affects the muscles and causes many of the underlying problems, particularly in growing lambs.

I'd never found a really effective way of getting a regular supply of the two minerals into the lambs at the time that they most needed it. Oral dosing helped both but only for a short time. An injection of the appropriate vitamins produced a longer-lived cure for cobalt problems, but still left the selenium problem unresolved.

I'd seen a few deficient lambs in the past few weeks. So, for the benefit of my flock, I was willing to listen to what this rep had to say, even if I did so with a rather sceptical ear. Judging by his youth and slightly nervous appearance, he was new to the job. As he went into his sales pitch, I understood why he felt so unsure. It turned out his company specialised in buckets. Not just any old buckets, but a range of pails containing a magical mixture of syrupy molasses combined with minerals supposedly guaranteed to cure a sheep of a myriad complaints.

I knew about these supplement buckets, but had never been persuaded. To treat a flock, a farmer would have to buy fifty or so of the £8 buckets. I remember a vet telling me that I'd be better off spending the £400 or £500 on a good night out. 'At least then you'd get something for your money,' he said. I'd come to regard them in the same way. To me they were nothing more than sweeties for sheep.

My scepticism should have been obvious. As we struck up a conversation, I told him that the only truly effective cure I'd discovered was putting the lambs 'out to keep' on the better soil of the inland dairy farms. I also expressed my doubts about the most widely accepted treatment on the market, a very expensive 'bolus', a large, bullet-like pill lodged in the sheep's stomach where it then slowly dissolves, releasing the required levels of minerals. I'd tried it once, treating half the lambs in a field, then marking them. Two months later there was no discernible difference between the treated lambs and the other half of the flock.

'This is a far better product, sir,' the young man assured me, seemingly encouraged by the mere fact I was even exchanging opinions with him. He'd probably been used to being dismissed with a curt, 'Not today, thank you'. We chatted for a few more minutes before he left. Even though I'd made my lack of interest plain, he clearly climbed back into his car feeling I could be won over.

The following morning at eight o'clock he reappeared in the yard. This time he was accompanied by a particularly self-important-looking chap who introduced himself as the area manager for the company. I wasn't best pleased. I thought I'd been more than generous with my time the day before, but had made my lack of interest plain.

I made it even more obvious this morning, but there was clearly some part of my repeated 'no's that they failed to understand. The area manager was obviously intent on demonstrating the art of closing a sale to his new apprentice. And as far as he was concerned, I was still there to be won over. The senior partner set off into his own sales patter, even more polished than that of his eager underling. But by now I was getting so irritated, I wouldn't have bought half-priced gold dust from either of them.

Clearly the area manager was used to fielding such negativity and had developed some kind of immunity to it. He must have been a snake-oil salesman in the Old West in a previous life, I thought later. Either that or he was deaf. Eventually, however, he could remain in his state of denial no longer, and clearly sensed the prospects of a sale were disappearing fast. So too was his reputation in the eyes of his junior. When I delivered my umpteenth shake of the head and weary, 'I'm not interested', the elder salesman resorted to his most desperate sales ploy yet. Reaching into the back of his car, he produced a sample product and gave it a loving rub, as if to allow its quality to shine through.

'It really is a lovely bucket,' he said, with a look of pathos in his eyes.

A few moments later master and apprentice were shuffling back to their Ford, doing their best to maintain the air of confidence upon which their professional lives clearly depended.

'That must be the worst job in the world,' I thought to myself.

When the phone rang in the middle of a quiet family Sunday, Debbie and I immediately exchanged ominous looks.

'What's the betting that's a walker who thinks they've spotted a problem with a sheep?' I said as I went to pick up the call.

The voice on the other end of the line was that of a man from the village, a regular Morte Point walker whom I knew. He explained the bad news in a voice that was clearly distressed.

'I've just watched a big black dog attack some of your sheep,' he said. 'The owner was trying to stop it but it had one of them right down.'

I can't be with the sheep all the time. The image of the ever-present shepherd, sitting under a tree tending to his flock at all times of the day and night, belongs back in the Bible. Instead I sometimes have to rely on others to let me know if a problem arises in my absence. Well-intentioned though they are, holidaymakers can't always be the best judges of what is or isn't a dangerous situation. So while the news was alarming, I was relieved it had been delivered by a local man. This meant I could be reasonably sure he wasn't going to be overreacting to something.

Grabbing a crook and taking only Swift from the kennel, I headed out to Morte Point as fast as I could. There was another reason I had been relieved the caller was a local. In the past I'd spent many a frustrating hour trying to make sense of a visitor's directions. Once a holidaymaker helpfully informed me a sheep was stuck 'where the cliff meets the sea', which narrowed it down to two and a half miles of meandering coastline. Today's caller couldn't have been much more precise with his directions. He told me the attack had happened a few hundred yards west of Greystone Gut.

I arrived there to find half a dozen people gathered on the lower side of the coastal path. It took me several minutes to walk down through the bracken on the steep north side of the headland, but once I made it down the reason for the small crowd was immediately obvious. A distraught-looking woman was restraining a large black dog, possibly an Alsatian cross. A few yards away a mutilated lamb lay gasping pathetically on its side in the long grass.

'He just ran off, there was nothing I could do,' the woman sobbed, looking down at her dog, giving him a reproachful punishing jerk on the chain as she spoke. The dog was completely oblivious to the problems it had caused, and looked down towards the stricken lamb as if intent on resuming what it obviously considered a legitimate game.

I bent down to inspect the lamb. Its injuries were far worse than they'd looked from a distance. The dog was powerfully built and had bitten deep into its neck. It had also torn through its belly, causing a wound through which the lamb's intestine was protruding. The poor creature now lay motionless in shock.

'Will he be all right? Can you take him to the vet? I'll pay for any bills,' the owner said, her state becoming if anything even more desperate as she saw the look of concern on my face.

I didn't answer. I was sure that she knew that the lamb was beyond help.

I carried it as gently as I could back up towards the Land Rover, where I lay it down once more. I took my knife from my pocket. Its suffering was over in a second.

I returned briskly to the scene. Two other walkers had told me another lamb had been chased by the dog. They were now standing peering cautiously over the cliff edge. 'It's down here somewhere,' one of them said.

The cliff wasn't vertical, but was pretty close to it. By edging out on a small rock promontory, I was able to look back, in the direction from where I could now hear bleating. I spotted the lamb apparently unharmed, stuck on a grassy gully that ran steeply to the rocks below. It was in no immediate danger, but a ten-foot section of cliff directly above meant it was unable to climb back up. Stranded, all it could now do was issue its piercing distress call.

I climbed back up the slope and spoke to the lady with the dog. There was no point in me remonstrating with her, she was already as upset as she could be. She gave me the address of her holiday accommodation, and we agreed that I would call round to discuss compensation. She walked homeward, with the dog still held tightly at her side. Each year I lose three or four sheep to out-of-control dogs. I couldn't help but think how much better it would have been for sheep, shepherd, dog and owner if this one had been restrained from the start.

I returned to the Land Rover once more, and this time came back with Swift and my crook. From my brief assessment, I judged that while I wouldn't be able to get to the lamb, Swift probably could.

We walked a little way along the edge, to where a track led down to the rocks. The tide was well out, and we soon picked our way through the great slabs of sea-worn slate, until we were at the foot of the gully directly below the lamb. Looking from below, it was now obvious that the climb directly up to the lamb was not as easy as it had appeared from above. However, a thirty-foot climb to the left looked as though it was possible. With the constant bleating from above echoing in the gully, Swift soon spotted her target and climbed willingly on in front of me. She stopped only once, unsure of herself as her claws struggled to grip on to the smooth rock face.

It took only a couple of minutes to find a ledge eighteen inches wide, just above the stricken lamb. From here I'd anticipated being able to almost reach it, but again the angle from above had been deceptive, and it was now clear that I was still some way away. This was where I needed Swift. The lamb's only desire so far had been to climb upwards to where its mother could clearly be heard. But below was a steep loose scree. I decided that if Swift could force it downwards, it could nearly slide the next ten feet. From there it would be comparatively easy for me to climb and grab it.

But like all the best-laid plans, this one didn't work out as I had hoped. With only a couple of words of command, Swift climbed down to face the lamb. Perhaps out of fear, or just purely its instinct to return upwards to the comfort of its mother, the lamb stared directly into Swift's eyes and froze. The eye of a collie is the key to its control over sheep. A dog with a powerful eye seems to only have to look at a sheep for the animal to know it's beaten. A weak-eyed dog, on the other hand, will soon be found out and challenged. The strength of Swift's eye held the lamb but still it refused to turn downhill as I needed it to do. To ask Swift to make a sudden movement would risk the lamb leaping. So I resorted to giving her a flanking command, placing her in front of the lamb. But still the lamb remained glued to the spot.

I climbed back down to the rocks to a position directly below the lamb and called Swift towards me. She came to look over the edge, about ten feet above where I now stood. As she tried to stop, she started skidding down the steep scree in the way I'd hoped the lamb would have done. In a shower of earth and stones, I caught hold of her as she arrived, somewhat faster than she had anticipated.

Plan A had failed and I didn't have a Plan B, but I wasn't ready to give up just yet. I'd noticed the rope from some fishermen's nets washed up among the rocks a short distance away. I walked down to collect it, then spent a couple of minutes cutting free a few short lengths as best I could. I then tied them together to make a length of about twelve feet. I tied one end of it to the crook of my stick then made a loop, which passed back through itself, at the other end. A minute or two later I was back on the ledge above the lamb, Swift this time watching from a distance.

Slowly I lowered my improvised lasso over the edge, until the loop dangled a few feet from the lamb's head, swaying with the movements of my arm. The next few minutes were a little like one of those fairground games, where you try to capture a duck with a hopelessly unsuitable grab. Fortunately, however, the lamb was still reluctant to move, and – with a speed that surprised me – I managed to loop the rope over the startled animal's head. With a minimum of ceremony, I pulled the loop tight, then hauled the struggling lamb back up to me. Ten minutes later, having carried the lamb back to level ground with some difficulty, I let it go. It ran off in the direction of its bleating mother, a hundred yards away.

Today could have been worse, I consoled myself, as the reunited ewe and lamb trotted off along the footpath together.

When I saw another unfamiliar car scrunching its way down the gravel drive to the yard, I felt myself bristling immediately. The attack on the lamb hadn't left me in the best of spirits and if this turned out to be another rep I was ready to give him a real earful.

It was obvious immediately that the elderly gentleman who climbed stiffly from the car wasn't here to sell me anything. 'Good morning. Hope you don't mind me disturbing you – are you the owner or the tenant?' he said, extending a hand. 'My name's Lawrence Moore. I've only been back here once in the last fifty years. I'm down here on holiday and couldn't help myself having a look.'

At first I assumed he was someone who'd lived or worked on the farm, but he soon revealed himself as someone altogether more significant in the farm's history.

'I was the co-pilot of the Lancaster bomber that crashed over there in 1945,' he said, pointing to the slopes of the field in front of the farm.

Ever since arriving at Borough Farm, I'd been aware of the story of how the small pit at the crest of the hill had got there. I'd never been entirely sure whether it was a rumour, but here was hard, human evidence that it was true.

Lawrence must have been in his seventies. He walked slowly and slightly awkwardly. As I accompanied him out to the pit, he began recounting the terrible events that had created this scar on the landscape.

It had been in December 1945, a few months after the end of the war, that he had been co-piloting a Lancaster bomber on its way out to North Africa on a mission to bring prisoners of war back to England. The crew had been only a few miles from land when first one engine then the other failed. 'My pilot had turned inland but we couldn't make it to a landing strip,' Lawrence said. 'We were losing height fast as we came up the valley and crashed into the hill, right here.'

Miraculously the plane's full fuel tanks hadn't ignited. The plane had been heading straight for the farmhouse when it had come down, so goodness knows what damage it might have caused. The pilot didn't survive the crash, but the other five members of the crew did, including Lawrence.

'I was in hospital for a year. Lost a leg,' he said. For a moment we stood there quietly surveying the scene, Lawrence wrapped in his memories.

'That's where I came round,' he said, pointing to a spot in the middle of the field. 'Someone must have pulled me clear.'

I invited him back to the house for a cup of tea. As we sat in the kitchen, he reached inside his jacket and produced an old but beautifully preserved photograph.

'This is me, before the crash,' he said.

In full RAF uniform, he looked the dashing epitome of a handsome young airman, aged just nineteen.

'Seems like only yesterday,' he said, wistfully.

Then he was on his way, leaving with a promise to pop in if he was down in this part of the world again. His car was soon easing its way back up the drive.

As I got back to work, I couldn't help dwelling on the events my unexpected visitor had described to me. Until today I'd looked at the farm and Borough Valley as something of a haven, a benign place, almost disconnected from the realities of the outside world. Yet in these fields one life had ended and another changed for ever. From now on I wouldn't be able to see the familiar landscape of the farm in the same light. More importantly, however, Lawrence had made me reassess something else. If my greatest headaches were dealing with unsolicited visits from salesmen and losing the odd lamb to a badly behaved dog, then I didn't have any real headaches at all.

CHAPTER TWENTY-FIVE

New Dogs, Old Tricks

The faint air of scepticism with which my four 'students' greeted me as I arrived in the field next to the house didn't do much to boost my confidence. Nor indeed did the motley quartet of dogs standing alongside them. 'What have I let myself in for here?' I thought to myself.

The reaction to my success with Greg and Swift at the English National had been pretty low key. But a couple of weeks afterwards I'd got an unexpected phone call from a lady called Caroline at a local horticultural and agricultural college. 'Congratulations on doing so well in Exeter. I was wondering if you'd be interested in teaching others how to train their sheepdogs?' she'd asked me. 'We'd pay you, of course.'

On the telephone, Caroline sounded an enthusiastic and determined character. She'd made it her mission in life to help teach the farmers of the South West new skills, as well as improve on the ones they already had. Her courses ranged from shearing classes to tuition on crop spraying. It was a tall order. Farmers tend to be strong-minded, independent people. The idea of training would seem alien to most of them – despite the fact that many of the courses Caroline ran were now compulsory to comply with the ever-changing agricultural legislation. But I felt she was fighting for a good cause, particularly on the importance of training sheepdogs. To me the benefits of working sheepdogs were obvious. Apart from anything else sheep themselves have a far less stressful life if worked by a skilled shepherd and dog. But it was also one of the many traditional shepherding skills in danger of dying out. So before I knew it I was agreeing to spend the next six months teaching the rudiments of sheepdog handling to a group of local farmers. The four people now standing in my field were my first intake of students.

We made our brief introductions, and I asked each one in turn about their respective dogs. My initial feelings about the dogs were – I have to confess – less than positive. The one that concerned me the most was a brown short-haired kelpie called Rusty that belonged to a younger farmer, Mike. Kelpies have been imported from the southern hemisphere and are a very different type of stock dog to the border collie. In Australia they are invaluable for working sheep yards and driving vast flocks large distances in a hot draining climate. But I had no experience of working with – let alone training – such dogs. Judging by the way Rusty was straining at the end of the lead, it was going to be a steep learning curve.

The second dog to catch my eye was Rex, a strong-looking black and white collie with a heavy coat. He belonged to a young dairy farmer, Robert. Of the four pupils, Robert seemed the happiest to be here. He lived and worked with his ageing grandfather and I quickly sensed he was glad of some different company.

My only female student was Jane. She and her husband had farmed in the area for many years. Her shy, Devonian nature couldn't have presented a sharper contrast to her rather enthusiastic-looking brown and black collie, Tod. From the moment he arrived in the field Tod had been intent on playing or fighting with the rest of his classmates, hardly encouraging behaviour.

The fourth and final member of the group was a neighbour, Robin. He was accompanied by a rather unusual-looking brown collie called Rosie. I knew Robin quite well. His family were prominent in the local farming community and I'd worked for him when we first moved to Mortehoe. Even then it had been apparent that his sheepdog handling abilities hadn't been his strongest asset. I was impressed that he'd decided to do something to improve the situation, if a little nervous at the prospect of being his new mentor.

Farmers like these four were a rarity. Most, in my experience, were full of excuses as to why their dogs weren't working under proper control. Some deflected the argument with statements like 'You only have one good dog in your life' or 'His great-grandfather was the best dog you'll ever see'. Others muttered cod wisdom like 'A good dog trains itself' or 'I don't have time to train a dog'. To my mind the latter was the least convincing argument of the lot. Whenever I heard it I had to bite my tongue to stop myself saying, 'I don't have time not to train a dog.' The countless hours that my dogs save me, to say nothing of the miles of walking, are more than ample reward for the time spent in the training field.

Apart from anything else, this lack of interest in sheepdog training has always struck me as ironic in the extreme. North Devon is one of the biggest stock producing areas in the country. Its farmers take great pride in producing sheep and cattle of the highest quality and spend considerable sums of money investing in prime breeding stock in order to do so. Yet at the same time they are reluctant to invest the relatively minor amounts of time or money needed to produce a good working sheepdog, a companion that will serve them every day for ten years, with luck. It made no sense whatsoever to me. It was another reason why I was glad to be making a contribution with Caroline.

For the purposes of today's lesson, I had brought ten sheep into the smallest paddock on the farm. They were grazing calmly as I began the session by asking each of the students in turn to show me how their dogs reacted to sheep. Mike went first with Rusty. My worst fears were at once realised. With the bounding energy of a caged lion released for the first time, Rusty lunged into the middle of the sheep. The startled animals bolted, taking refuge in the corner of the paddock where Rusty stood barking at them, his tail raised in excitement. Mike retrieved the dog rather apologetically.

Next to go was Robert with Rex. At least here I drew some mild comfort. There's only so much that can be trained into a dog. As the trainer you are to some extent reliant on its natural herding abilities. Unlike the rest of the class, Rex at least seemed to have some of the right genes. The way that he started to herd the sheep towards Mike gave me hope that we had something with which to work. It was going to take a lot of hard work, but I felt he could be a useful 'farm dog'.

Third to go was the energetic Tod. I'd been watching him. Jane had been trying to calm him down by stroking him reassuringly. The two obviously had a close bond. She carried on talking to him as they walked forward.

'My 'usband says 'e'll be nowt but a zooner dog,' Jane said to me as she readied him for his run.

I felt a brief moment of panic. I was meant to be the expert, but I had no idea what she was talking about.

'A what?' I asked, realising I had no choice but to admit my ignorance.

'You know, a zooner dog,' she said. ''E'd zooner bide at home, than go out an' do a day's work!'

She'd only tried Tod with a flock of sheep a couple of times. ''E wasn't much interested,' she said. It was soon obvious he hadn't been given the chance to get interested. No sooner had she let Tod off the leash than she was trying to call him back again. ''Ere, Tod. No, 'ere, Tod. Come 'ere,' she scolded him.

Sheepdog training should be a positive experience. Jane's panic was sending nothing but negative messages to Tod. The more she shouted at him, the more wound up he became. This, I could see, was going to be the first problem we'd have to overcome.

A minute or two later, after a brief encounter with the sheep, in which Tod had only shown a momentary interest, Jane re-leashed him and walked cautiously back towards me. 'What do you think?' she asked. 'Will 'e make it?'

'It's early days yet,' I said as diplomatically as I could. Jane seemed to sense my reservations.

Robin was the last to show me what his dog could do. Within moments of Rosie setting off I realised here too I was going to be working with a dog of limited potential. There are certain traits that are common to almost all good working dogs, the most obvious of which is the way the tail is tucked down low. It's a sign of concentration, and from the way Rosie's tail waved around in the breeze like a flag, it was clear this was something she was going to have a problem with. Sure enough, as she ran in the vague direction of the sheep she was immediately distracted by a low-flying swallow. Rosie pursued the bird with vigour and only gave up when she nearly collided with the fence. Robin called her back then walked over to where the rest of the class had gathered by the gate.

'Well,' he said, as if pleased that I'd landed myself with such a challenging collection of ill-disciplined canines. 'There wouldn't be any point in us coming if they were all perfect to start with.'

CHAPTER TWENTY-SIX

Profits and Losses

'Thirty-four . . . thirty-four, four twenty . . . four sixty . . . four eighty . . . thirty-five.' As the auctioneer scanned the gathering for improved offers on my sheep, it was hard to know which was rising faster – the price or my pulse rate. I usually have the uncanny knack of selling at market in the very week that prices collapse, and buying just as they peak. But this morning it looked as if the habit of a lifetime was going to be broken. I nearly had to pinch myself.

I'd almost not brought the forty old 'draft' ewes in for the sale at Blackmoor Gate. It was rather early in the season to be selling off these older ewes who were now well past breeding, and when I'd discovered a puncture in the tyre of the sheep-trailer early that morning I'd almost abandoned the trip altogether. The thought of paying for a livestock haulier didn't appeal much, but I'd made the call and a lorry had arrived just in time to get us to the market and the eleven o'clock start.

As the bids for the first batch of half a dozen of the better ewes kept coming in, I was glad I had persisted.

These markets represent the only real way that a farmer has of determining the true value of his stock; however, each sale takes up a morning to attend, time that is difficult to spare. This morning it felt like time well invested.

The price of sheep varies wildly. What they will fetch from one week, month or year to the next is anyone's guess, and you can't even rely on them going up historically. For instance, when I first started out, twenty years ago, good breeding lambs might have made £45 each. Four years ago, they might have got £60. But two years ago, they'd have struggled to fetch £35. The price of old ewes like those I was selling today varied just as much. With variations like this in the market, it's almost impossible to make plans from year to year, although it has to be said we sheep farmers don't help ourselves sometimes.

Our industry is made up of thousands of small individual businesses, none of which, it seems, is able to look at the big picture. When prices crash and profits disappear, we all try to compensate by keeping a few more sheep. It's a strategy that is inherently self-defeating, because the following season there are even more lambs on the market and the simple laws of supply and demand ensure prices suffer again.

It was clear that over-supply wasn't a problem today. By the time the auctioneer had finished prising out new bids, the price stood at thirty-five pounds eighty pence each. It was a very good price, nearly twice as much as the same sheep made twelve months earlier. As the gavel came down, a murmur went round the fifty or so farmers crowded around the ring. Everyone there could see these sheep were well past their prime, and was probably wondering how much they were going to make in a month's time in September, when Blackmoor Gate held the main market for prime breeding sheep.

The price for the next batch did nothing to puncture the sense of optimism bubbling away. The next pen of ewes to enter the ring were rather less plump than the first batch. Once again, however, the buyers' enthusiasm was surprising.

Agricultural auctioneers don't waste their time in getting through the lots. 'Sold at twenty-seven,' he was soon announcing with another bang of his gavel. At only eight pounds less than the first pen, I couldn't hold back a faint smile. Before long the last five of my ewes were waddling into the ring. They were elderly customers and, although they were bright and healthy, I couldn't help feeling a little embarrassed. They'd failed to thrive on the rich summer grazing, and would certainly fare no better as the autumn approached.

A few wise heads shook as the auctioneer pronounced them sold at nearly ten pounds each. I knew too I'd done well to get that much. I headed for home feeling rather satisfied with the morning's work, and wishing that I'd bought the rest of the season's 'draft' ewes in for such an unusually buoyant trade.

'Who knows, I might be able to afford to get third gear fixed on the Land Rover,' I thought to myself, as I ground my way through the gearbox once again.

I was feeling rather pleased with myself when I arrived in the kitchen for an early lunch.

The morning's activities meant I headed out for the daily check on the sheep a little later than usual. I set off intent on giving them no more than a fleeting once-over this afternoon. With all the flock treated against the major problem of the late summer, fly strike, there shouldn't have been too many problems.

But as I cast my eye around the recently weaned lambs in the highest field on the farm, Borough Heights, the sight of a woolly lump by the gate filled me with an immediate sense of foreboding. On further investigation, I was greeted by a scene that was as sickening as it was incredible. Two lambs lay dead at the foot of the gate, each with its neck wedged in the V-shape created where the horizontal bars of the gate met the diagonal brace. Quite why the two lambs had both decided to push their heads through at the same time was beyond me, but having got their heads stuck they had managed to cross over one another, securing the grip so tightly as to make the death hold inescapable. Now blown up like balloons in the late August sun, the two lay grotesquely twisted together, victims of their own stupidity.

Suicidal sheep are no rarity – indeed, every shepherd can tell a story of the demise of an animal in a manner that defies belief. No good stockmen would ever leave a loop of string in a lambing shed. An inquisitive lamb will invariably pass its head through the loop, then strangle itself in its attempts to get free. Ponds, particularly if covered with a layer of weed, will frequently entrap a sheep that believes it can walk on water, and the temptation of choking on a plastic bag left in a field can be too much to resist for many.

A lot of farmers suspect that when old age finally takes its toll, sheep often deliberately pass away somewhere visible just to embarrass their owner. A good friend once told me of the only field he had on his farm that had a public footpath running through it. 'I've only lost three sheep this spring,' he said. 'And all three of 'em died right on the footpath.'

Although by keeping a young flock my mortality rates tended to be below the average of four per cent a year, inevitably deaths occur. Out on Morte Point on one occasion I retrieved a deceased ewe that had chosen a spot right in front of a bench at a particularly fine vantage point – quite literally preventing anyone from sitting and admiring the view.

Even by these standards, however, the two lambs in front of me represented a new level of misfortune. I'd accepted long ago there was nothing I could do to prevent things like this happening. All I could hope was that, with a little luck, I would occasionally find myself in the right place at the right time. On this occasion I hadn't been there.

I loaded the two carcasses into the trailer for disposal, cursing their stupidity as well as the financial loss they represented. Suddenly the success of the morning's market didn't seem so sweet after all.

As August wore on, so too did the repetitive routine that is sheep husbandry. Today I'd had another four hundred or so to dock and put through the footbath and had chosen Greg and Ernie to help me. Greg was hugely experienced at working the sheep through the pens and had needed few words from me. This meant I'd been able to keep talking to Ernie, constantly reminding him that I was in charge.

I felt I'd reached a key point in Ernie's training. I could now take him to work almost certain that he was going to be of assistance. This wasn't something I'd ever been able to say about him before. But now he was nineteen months, I really felt he had turned a corner. He certainly had been working well this morning.

Greg's opinion of Ernie was obviously changing as well. Up until now Greg had, it seemed, regarded him as no more than a pup, whom he could ignore for the most part. But as Ernie was maturing, Greg had reappraised him. There was still no doubt who was top dog. On more than one occasion this morning, Greg had decided that if Ernie ignored my instructions repeatedly, then it was down to him to instil a little discipline. If my rather desperate sounding 'Go out, Ernie' was ignored, Greg would run over, his hackles raised, and growl at the impudent adolescent. Ernie would cower on the floor, put in his place by the senior member of the farm's pack.

With all four hundred lambs treated, the final job today was to help drive them back to the fields. As ever they swirled in a great mass around the yard, none of them inclined to take the lead and head off up the drive. It took fully five minutes before one bright spark eventually spotted the required exit. As the flock moved off in the right direction, the dogs tucked in behind the last sheep.

In a narrow lane, with high-sided banks, four hundred lambs can stretch themselves out over a couple of hundred yards. Some will climb at the walls nibbling at fresh shoots, while others tend to run on ahead. As we neared our destination, I was relying on the lambs to make a 90-degree right turn into the field, as they always did. But instead, as I came over the brow of the hill on the lane, I could see the lead group of lambs had gone past the gateway and were running headlong towards the main road at the top of the farm lane a hundred yards away.

Overtaking or 'passing' sheep on a lane is a skill which comes naturally to some dogs. Swift is an expert at it, but she was at the farm shut in her kennel. Greg wasn't bad at it but, with the lambs tightly packed in the lane, I knew he would struggle to force his way past them before they got to the road. I'd never found myself in a position where I'd had to ask Ernie to do something like this. But now I had no option but to try.

With a command to both dogs, they set off to head the flock. Soon they were trying to forge their way between the bank and the lambs. Ernie quickly disappeared into the chaos but, as I'd suspected, Greg was soon struggling to find a way past.

My fear now was that the entire flock was destined to reach the main road before the dogs. The prospect of four hundred sheep wandering the roads during the busiest time of the year wasn't an appetising one. But to my amazement, I saw the lead lambs start to turn with a few yards to spare. As they did so I saw the black flash of Ernie appear at the head of them.

He'd done brilliantly to get there, but his job wasn't over quite yet.

A flock running at speed down a narrow lane can take a lot of stopping. Although the first sheep had been turned, the back of the flock was still running up the lane, pushing against the lambs ahead of them. The pressure sent the leading lambs forward once more towards Ernie. But as they tried to break past, he was equal to them all. He stood wilfully in the centre of the drive, his strong eyes flicking from one lamb to another, darting from side to side with lightning speed as he prevented any of the flock from passing him.

Eventually Greg forced his way past the last of the flock and the lambs began their retreat down the drive. Having been at the back of the flock on the way up, I was now in front, and at last perfectly positioned to direct the lambs through the gateway on the way back down.

I called Greg and Ernie back as they followed the last animal through the gates. Greg stood still looking at me and wagging his tail. Ernie on the other hand was still watching the flock intently. As they dispersed in the field, it was as if he was hoping he might be sent to gather them once more.

'Come here, Ernie,' I called to him. He came over and sat at my feet, giving me that trusting look of his once more. I rubbed his ears affectionately. He'd acquitted himself well in the most difficult and potentially serious situation of his working life so far. A small part of the potential he'd shown since his earliest days had been fulfilled.

As I made a fuss of him, I couldn't help but think back to where we'd been just five months earlier. Back then I'd despaired – now I felt part of his apprenticeship was over. Perhaps his days in the last chance saloon were over.

Autumn

CHAPTER TWENTY-SEVEN

No Money for Good Sheep

The morning had a distinctive early September feeling to it. A damp, penetrating mist hung on the chill breeze, obscuring Borough Valley below me and reducing visibility to no more than a hundred yards even on the higher ground. Passing the pond at the edge of the fields, it was clear I wasn't the only one feeling my way through the shrouded landscape. Just ahead, a stoat dived under the hollow of a half-rotted log, then poked its head out again, nervously scanning the scene. The thought there might be danger at anything above knee level clearly didn't occur to him. Happy the coast was clear, he ran for his next safe haven, seemingly oblivious to the fact he had scurried across my boot along the way.

It was an early start this morning, preparing for Blackmoor Gate Market, where I was taking a hundred breeding lambs. I had a lorry arriving to collect them around eight a.m., and had set off with just Fern to get the lambs in.

Fern had undoubtedly improved steadily over the summer, although every now and again the sensitive side of her nature still showed itself. I'd see a look of indecision on her face, or the now familiar wag of the tail, followed by a dropping to the floor with her ears tucked down, when she was not sure of what she was being asked to do. But she'd impressed me in other respects, with her stamina in particular. On the hottest days, when some of the other dogs were flagging, Fern had kept going, displaying what seemed like limitless energy. Her outstanding hearing ability meant she picked up every command and she was now working over increasingly long distances.

My faith in her this morning again proved well founded. At my command she was off in a flash and soon bringing a hundred lambs out of the mist, across the field and towards the sheep-pens.

These Suffolk-cross Mule ewe-lambs were the only lambs that I sold each year as breeding sheep, and they had been receiving preferential treatment in the past few weeks. Since weaning they had been kept on the best grass, given an extra worm dose and been foot-bathed at every opportunity. Appearance is everything when selling breeding lambs, and 'clean-headed' sheep always sell well. So I had also spent time over the last week, removing any 'unfashionable' wool from the top of the heads and cheeks.

I couldn't help thinking that the lambs looked a picture as they ran into the yard, a cloud of hot breath rising above them, their coats sparkling white from the covering of morning mist. At six months old, they were a good even-sized flock, each weighing in excess of forty kilograms – over half their adult weight – and each with a jet-black face.

At market the lambs would be sold in pens of ten. To maximise the price, they needed to be sorted so that each pen contained animals of a similar size and appearance. Ideally they should look like peas in a pod. I spent nearly an hour arranging the lambs by size, face markings and coat, changing my mind at least half a dozen times with each one. By the time the lorry came clattering down the farm drive, I felt I'd sorted them pretty well, and was feeling quietly confident of a good sale.

Blackmoor Gate is the only livestock market on the west of Exmoor. So given that Exmoor has supposedly the highest density of sheep in the country, it is not surprising it attracts a lot of business. But with September traditionally the month for farmers to buy and sell the year's breeding stock, today's market was particularly busy, with as many as ten thousand sheep, and buyers from far and wide.

As my lorry arrived and waited its turn to unload, the mist was thinning into a light, hazy rain and I could see that the market was already a hive of sometimes chaotic activity. Groups of sheep were being driven in all directions. A gate had been left open and three ewes had escaped into the car-park where they had disappeared among the already dense ranks of vehicles. Half a dozen farmers and market drovers were engaged in frantic pursuit.

Everywhere else, it seemed, buyers and sellers were making better progress with their morning's preparations. Ever since the market began a hundred years ago, Blackmoor Gate has been a showcase for the quality of stock produced by Exmoor farmers. Near the lorry, a pen of beautiful, strong-looking Exmoor Mule lambs, being sorted by a group of farmers, were testimony to the generations of knowledge that had been applied to raising them. When they'd finished picking through the sheep, their owners stood there for a brief moment, a justified look of pride on their faces.

Elsewhere one of the market clerks was already walking along the pens, counting the sheep in each, then recording the numbers on the auctioneer's sheet she was protecting from the drizzle under her wax jacket. A few potential buyers were already weighing up the quality of the day's stock. I saw a thickset young man in a Stetson climb into a pen then methodically feel the lambs along their backs, before marking something in his catalogue and moving on.

He was one of the few youthful faces around. Admittedly, on farms worked by two or three generations of the same family, it tended to be the older members who took stock to market. But even allowing for that, the market was overwhelmingly dominated by white-haired figures, most of them well into their sixties. Like me, I suspect, many of these men had sensed long ago that sheep farming would cease to be a viable way of life soon. They had probably encouraged their children into other careers away from agriculture, thus slowly draining away the inherited knowledge that has shaped the countryside for generations. Whatever the rights and wrongs of the problems that have afflicted British farming in recent years, I think most people still believe we should remain capable as a nation of feeding ourselves. No industry can survive without young blood coming through, however, and markets like Blackmoor Gate only serve to prove that unless something is done to redress the balance, ours is going to literally die out. It was a saddening thought.

It was fully a quarter of an hour before my lorry was brought forward for unloading. According to the catalogue my sheep were down to be sold on the wooden pens in the field adjacent to the main sales area, where hundreds of wooden hurdles had been erected to form six rows of pens. The downside was that the wooden gates were no more than knee high, and I knew from experience that I was probably going to spend the best part of the day trying to prevent the lambs from jumping out.

With a little persuasion the lambs ran from the lorry into the pens. As I got ready to start sorting them I spotted Robin, minus the brown dog he had brought along to the training class a fortnight earlier.

'Give me a hand to sort these lambs,' I called out. 'There's a cup of coffee in it for you.'

I'd put a small coloured crayon mark on each lamb of the same group, but it still took more than half an hour for Robin and me to organise them into ten pens of ten lambs.

'Can't think why you've got that lamb in that pen,' Robin said, as we both stood there looking over what I thought was a 'tidy run' of lambs. 'It's half the size of the rest.'

Eyeing the lambs again, I had to admit he was right.

'And that one there doesn't fit, it should be three pens up,' Robin went on, as if warming to his task. Annoyingly, he once more had a point.

'Is that all right for you, sir?' I asked, with heavy sarcasm after I'd acted on his advice, replacing the two relocated lambs with better-matched specimens.

I was actually quite pleased to have Robin's opinion. His family was one of the longest-standing farming families in North Devon. At the latest count, seventeen members of the current generation were farming locally. He was chairman of the farmers' consortium which had bought out Blackmoor Gate Market when its previous owners wanted to pull out, and in doing so had been instrumental in saving a highly valued asset of the local farming community.

'Come on then, looks like I owe you a coffee,' I said when we'd both satisfied ourselves we'd done all we could to match the pens.

The cafe, like the rest of the market, was what you might call traditional. It was little more than a small corrugated tin shed, with a concrete floor and no lighting. It would have been large enough for a dozen tables, but the shed also doubled up as the market workshop. Various pieces of wire, tools and fencing posts cluttered the first few yards by the entrance, which meant there were only eight tables and chairs squeezed in this morning.

Blackmoor Gate is more than a mere market; it is also a place where otherwise isolated farmers get together to talk shop. So despite its ramshackle appearance, the cafe was filled with a dozen or more farmers, sitting in small groups, chatting in low voices over steaming mugs of coffee.

'So what's the trade going to do today, Robin?' I asked him, handing him a mug of coffee. 'You're the man in the know.'

'Well, it won't be as dear as when you brought those draft ewes in last month,' he said with a knowing look. 'Someone told me he'd never seen sheep so expensive as those that chap Kennard from Mortehoe sold.'

News travels quickly among farmers, and Robin was as well placed as anyone locally to hear the gossip, but I was still a bit lost for words.

'It was about time I had some luck here,' I replied defensively. 'Anyway, how do you know about that?'

'Never you mind. But don't look so worried,' Robin said. 'You'll be all right with those lambs – especially now you've had someone with decent eyesight sort them for you.'

As I headed back towards my sheep, the sun was at last trying to break through and the whole scene looked brighter. The last few pens were being filled as the ringing of a bell heralded the start of the sale. An old tractor fired into action, pulling behind it a rickety trailer on to which a corrugated tin auctioneer's hut had been roped. Moments later the trailer stopped by the first pen of sheep to be sold, and the auctioneer's voice rang out searching for the first bid. I walked on back to my pens – it would be a couple of hours before they reached me.

I arrived back to find a slightly rounded grey-haired man, leaning heavily on his stick, casting a critical eye over my sheep, who – to my mild amazement – were still inside the pens as I'd left them ten minutes earlier.

'Morning. Looking for some ewe-lambs?' I said, seeming perhaps a little too jovial. He cast me a sideways glance and nodded but gave no more than a grunt. Undeterred, I dropped into the nearest thing to a sales patter that I possessed.

'All out of good North Country Mules,' I assured him. 'Regularly wormed, dipped and vaccinated, off the poor ground at Mortehoe.'

He looked at me again out of the corner of his eye.

'March born?' he grunted again, this time by way of a question.

Obviously the later born they were the more they would have to grow.

'No, April,' I replied, stretching the truth a little bit, 'Take them home and they'll grow like mushrooms,' I added, knowing every other vendor at the sale would use the same line during the course of the day.

'Why 'aven't you sorted them?' he asked bluntly. 'I shan't be biddin' on 'em sorted like that – look, 'e's got completely different skin to 'im!' Of the two lambs he had pointed out, one had a tightly woven fleece and the other a more open coat. It wasn't something I'd thought was important, but clearly everyone had an opinion. Apparently content at having put me in my place he limped away, his deep-lined scowl still intact, leaving me somewhat speechless.

Slightly sooner than I'd anticipated, I was relieved to see the tractor and rostrum at last heading down the row towards me.

All manner of factors affect the price at market: supply, demand, the abundance of grass around at the time, the price of finished lambs and breeding sheep at other markets. That's even before you consider the quality and type of lamb you are selling. The fact remains, however, that a lamb is only worth what someone is prepared to pay for it, so farmers are price takers not price makers. I'd watched several pens of lambs similar to mine sold, and it looked as if prices were pretty good, but there were still butterflies in my stomach. All sorts of thoughts buzzed around in my head. Perhaps the buyers had spent all of their money? Maybe everyone who'd come looking for Suffolk-cross Mule lambs had already found what they wanted?

The tractor stopped with the rostrum directly behind my pen and suddenly at least thirty farmers leant over the bars of the hurdles, those closest feeling the lambs and muttering quietly among themselves. I took up a position among the lambs.

'Suffolk-cross Mules from Mr Kennard,' boomed the auctioneer. 'Unsorted lambs, ladies and gentlemen . . . as they ran off the lorry.'

Before I could remonstrate, he was looking for his first bid.

'Forty pounds straight in.'

I looked around anxiously. I thought these sheep were worth fifty pounds a head, and knew that if he couldn't get an opening bid of forty on this, the best pen, then it was going to be a disappointing day. But my luck was in. The auctioneer spotted a bid of forty pounds from a ruddy-faced man with straggly grey hair, and we were off. Now there was another bid, this time from a local sheep dealer. He must have had a customer willing to pay a good price, because much to my delight he and the ruddy-faced man were soon locked in a bidding war. Jumping fifty pence with each bid the price was soon near the fifty-pound mark. I relaxed a little. Anything above that would be a bonus, as far as I was concerned. When the hammer fell at eighty pence over the fifty, I knew I'd got a good price.

The last thing to do in the middle of a sale is show satisfaction, however. It might give the impression that I'd settle for less. Instead I put on the most dissatisfied look I could manage, muttering phrases like 'given away', and 'It's no money for good sheep.'

By the time the auctioneer got to the fourth pen of lambs, which I knew were significantly smaller than the first, the price had slipped by three pounds. I caught his eye before he began.

'Are you selling, Mr Kennard?'

I tried to pull a disappointed sort of face, and looked to be deep in thought for a few seconds.

'More like being robbed,' I said. 'But yes, go on.'

The ruddy-faced man who'd by now missed out on the first three pens was still bidding. This time he was in luck. When the hammer fell on the fourth pen, he'd finally secured his first batch of the day at a price well over fifty pounds a head. In North Devon markets, a buyer will frequently be given the option of paying the same price for the remaining stock of the vendor from whom he or she is buying. The auctioneer clearly knew this man was interested in a reasonably large number of sheep.

'How many pens?' he asked.

The buyer scanned the remaining half a dozen pens of my lambs carefully, before nodding first to himself then the auctioneer.

'I'll take the lot,' he said.

It was all I could do to suppress a giveaway smile.

'I think we'd better make a start,' I told my 'students' as I watched the first fingers of a new sea mist working their way in, wraith-like, from the coast and up the valley. 'While we can still see something.'

It wasn't that I hadn't enjoyed my first tastes of training instruction, but the fortnightly classes did seem to be coming around rather quickly. Today's was the third, but – to my mind, at least – the most important because Caroline had come along to assess how we were getting on.

I'd only ever spoken to Caroline on the phone so I was curious to meet her face to face. In an Afghan coat, with her hair up in bunches, she looked like she'd be more at home at the Glastonbury Festival. She introduced herself before taking up a vantage point from where she was going to watch the morning's session. She'd arrived to discover my class reduced from four to three. I hadn't been too disappointed when Mike had rung the previous evening, saying that he hadn't had time to work with Rusty since our last meeting, and that with his complete lack of progress so far he thought it might be best if he dropped out of the course.

The morning air was cool, the overnight rain had left the lush grass wet underfoot, and the ground puggy for the first time in months. The small stream that flows through the bottom end of the paddock had dried in early June, but was now gushing noisily through a pipe in the dam of a small pond.

First to put her dog through its paces this morning was Jane. I asked her if she felt she was making progress, but she sounded none too confident.

'Sometimes 'e shows some interest in workin', but when 'e does, 'e don't seem to know quite what to do,' she said, looking doubtfully down at Tod. Tod in turn was looking hopefully towards the other canine pupils present, ears pricked and tail raised up over his back. His attention was anywhere but on the ten sheep that Greg had retrieved and was holding a few yards in front of us.

'Just let him run and keep encouraging him,' I told Jane. 'If he runs off and away from the sheep, just keep calling his name in an encouraging tone.'

A moment later Tod was off again. He was apparently now more sheep-motivated, and ran into the middle of the small flock with a playful bark which scattered them in all directions. As Greg tried his hardest to reunite the splintered groups I worked with Jane, showing her where to stand, where to move and how to best encourage Tod to hold the sheep together. Tod's interest in the job of herding sheep was now growing. No sooner had Greg brought the straying animals back together than Tod ran straight into them once more, this time half mounting one of the unfortunate animals.

Training a sheepdog for the first time is a difficult skill to master: moving to the correct position, encouraging your dog, while trying to stop it making the inevitable mistakes that send the sheep out of control. It all happens fast, but it's all so much easier if him with which you are working has an inherent ability. Watching Jane with Tod I found difficult, almost heart-breaking. Jane was so obviously attached to him, as he was to her. It was also obvious that she was determined to train a sheepdog that she could enjoy working with, and of which she could be proud. I just couldn't see how it was going to be possible for Tod to fulfil the role. Jane walked back over to me, with Tod once more on his lead. I tried desperately to think of something positive to say about his performance, and to give some constructive advice, but it was difficult to find inspiration. From the quick glance that Caroline gave me, she obviously knew what I was thinking.

A few minutes later Robert was putting Rex through his paces. Now here I could see some improvement. Robert had put in a lot of time with Rex over the last couple of weeks, and both were obviously starting to enjoy their training sessions. Rex was beginning to understand that in order to control his sheep, he needed to give them room. He executed several nice runs, arcing widely around the sheep. Robert had taught him to stop on command and, although the work he was doing was basic, I felt reassured that Robert and Rex were making good progress. Perhaps my training was achieving something after all.

Robin confessed that he had been a bit short of time over the last couple of weeks. When he let Rosie loose it was obvious that she hadn't improved. Despite the attempts of both Robin and me to encourage and cajole her to take some notice of the sheep, it was in vain. Rosie just couldn't see any reason to involve the sheep in her run around the field – she was getting all the exercise she needed without them.

Calling her back with a few affectionate words of praise, Robin turned to me. 'I've got another dog, Penny, in the truck, shall I try her?'

I remembered Penny from the time I'd worked for Robin and calculated she must now be nearly nine years old. As Robin sent the somewhat tired-looking Penny around the sheep, the words old dogs and new tricks inevitably sprang to mind. It gave me an opportunity to give him a few hints on handling, nevertheless.

'Come bye, Penny, come bye,' he called to her as she made a rather sedate flank past the sheep. Robin looked at his hands, as if weighing up which way was which. 'Oh that's left, isn't it. Perhaps it's "away". Penny . . . try "away".' Dogs are dependent on clear, consistent commands. If Robin changed his mind like this all the time, it was no surprise Penny had never really mastered her left and right commands.

Robin returned to the rest of us, with Penny. At least I felt that I could help him understand where he might have gone wrong in the past. As the lesson drew to a close, Caroline rejoined us. She gave a very diplomatic talk, about how interesting it had been watching the three 'very different types of dog'. Caroline had previously expressed a desire to one day train a sheepdog of her own. I wondered if this morning had put her off.

'Not at all,' she said. 'It's made me even more determined to give it a go.'

Just as the mist threatened to engulf us completely, we made our way back to the house to deal with the required paperwork. Over a cup of coffee Jane asked how long the course was likely to go on. Caroline gave me a glance before replying. 'A little longer than I'd first envisaged,' she said, a smile creeping on to her face. I joined in the round of laughter that followed, hoping no one would detect the slightly hollow tone of mine.

The September rain brought with it a flush of autumn grass and welcome pickings for the flock. As I walked through the lower fields that afternoon, it wasn't just the ewes enjoying the late season bounty. A 'charm' of goldfinches was flitting between the plentiful seed heads of some thistles that lined the top of a bank.

With Gail, Greg and Swift I was heading for the fields furthest from the farm, where a flock of more than three hundred lambs were grazing. These lambs were the smallest of this year's crop and still needed to grow a good deal before they could finish. As I ran a routine eye over them today, there was still the odd scabby ear too, the telltale sign of cobalt deficiency. I'd hung containers of powdered minerals out on the fences but clearly they hadn't all been eating them. I'd need to treat them within the next week.

By late summer and early autumn, it's fairly clear whether it's been a good or bad farming year. As I turned towards the next field, I found myself taking stock once more.

So much that dictates the financial success or failure of the shepherding year is out of my control. All I can do is prepare the flock for tupping, manage the ewes through winter, work hard over lambing, and hope for a little luck along the way. The rest is up to the weather but, more importantly, the market. So the sense of relief I'd been feeling in the wake of Blackmoor Gate had been immense.

The final lambing tally had been around 1,250, roughly 1.6 lambs per ewe, not a spectacular percentage but good enough. I'd also sold a decent proportion of the remaining flock to my biggest customer, Lloyd Maunder. Only five hundred remained here at Borough and at Town Farm.

If these remaining lambs finished without the need to buy in too much feed, then sold as well as the first batches, then the business would probably survive intact for another twelve months. These were big ifs, of course, but that's the unpredictable nature of the business. That was all, realistically, we could hope for, but all in all it had been a fairly successful year.

My internal audit was interrupted by a piercing, screeching sound. Half a dozen lambs had obviously broken out earlier in the day and were now returning to the lower field via a hole in a fence. The high-pitched noise was being made by the wire as it was being pulled through a staple on a fence post. I'd already made a mental note to replace this section of fence in the coming winter, but had consigned it to the back of my mind.

This old field boundary wound its way along the steep edge of the woods for five hundred yards, rising and falling sharply, so the whole task would have to be undertaken by hand. With rock only a few inches below the surface, the fence posts would be extremely difficult to drive in. It wasn't something I was looking forward to.

By the time I'd reached the top of the farm once more, the light was fading. There was no sunset; instead the sky was now laden with heavy, black clouds and the promise of more rain.

My last job outdoors was to check on the ewes and rams, already being prepared for tupping in a month's time. The ewe flock were beginning to look in fine fettle. It's amazing how the appearance of a flock is suddenly altered by sorting the fit, the fat and the thin. It took me only a few minutes to check on them before heading for the upper fields where the rams should have been.

It was getting late and they'd wandered off from the feeding troughs where they would have gathered earlier. I sent Gail and Greg to find them. As the two dogs disappeared over the brow of the hill, it occurred to me for the first time that Greg was conceding a yard or two to his daughter. But what he lacked in raw speed these days, he still more than made up for in sheer brains. A minute later the rams duly appeared, under the guidance of Greg. In her haste, Gail had apparently got lost.

I was already committed to buying in a couple of rams, as I did every year. Two rams had died the previous winter and, despite feeding the others up for several weeks, a few more had begun to look a little the worse for wear. I doubted they'd be able to cope with the coming months' activity. I'd probably need another four rams – a large but unavoidable expense.

CHAPTER TWENTY-EIGHT

The International

The hot summer days were becoming just a distant memory as I began to prepare the ewe flock for tupping once again. The temperature had dropped dramatically, bringing with it more mists and occasional showers.

As I hung a new field gate this morning, however, my thoughts were elsewhere. I couldn't rid myself of the nagging anxiety at the back of my mind. It was only a week before the International sheepdog trials in Aberystwyth, seven days before I was going to do something I'd never in my wildest dreams imagined myself doing – representing my country at sport.

Old hands talk of sheepdogs being 'made sour' by too much training, so in the days and weeks since the National I'd tried not to overdo the practice for fear of boring Greg and Swift. A week earlier, however, I had entered them in the one and only brace competition of the year in South West England, at Brentor. I'd thought it would be a good opportunity to polish up their competitive skills and give us all a little confidence. It had done the exact opposite and I'd wished I'd never gone.

Competitors in other sports say this, I know, but sheepdog trialling really is a great leveller. Just when you think you are doing well, with a few good results behind you, things tend to go wrong. Of course there are plenty of explanations for this, not least the fact that you are dependent not just on one dog – or in my case, two – but also on half a dozen determinedly awkward sheep. At Brentor the sheep were cussed from the start and, as hard as we tried, no amount of effort from Greg and Swift, or instruction from myself, could push them in the right direction. The words of consolation I received from one or two other competitors did little to ease my concerns. The International was intended to be the ultimate challenge and was going to be on a course difficult enough to test the best sheepdogs and handlers in Britain and Ireland. Just as worrying to me, the audience would be the most critical that I was ever likely to compete before. We were going to have to do an awful lot better than that if we weren't to let the England team down.

I'd seen Andrew in the days following the setback. I suspected he was feeling a little responsible for the fact that I had qualified to run in Wales at all. I wasn't entirely surprised when he rang up mid-way through the morning.

'I've got a course set up for you to have a practice on,' he said. 'If you can do this one, you should be able to do anything they set you next week.'

I drove the short distance to Andrew's farm that evening. He certainly had found a course to test both me and the dogs. It took the dogs four hundred yards to the bottom of the field, through a gate, then up a long climb to the sheep at the top of a hill. This time things went rather well. Greg and Swift took it in their stride, and the ease with which they found their way, then controlled the sheep, left me feeling a bit more optimistic.

'Next week's course can't be any more difficult than that, can it?' I asked Andrew, trying not to sound desperate. He had been to several International finals before, and had far more idea than I did what to expect.

His muted response left me feeling less than reassured. 'We'll see,' he said.

Seven days later, words couldn't do justice to the level of nerves gripping me as I stood on a hillside outside Aberystwyth. In front of me, two or three thousand people filled two large grandstands. To judge from the accents I'd heard, people had come from as far afield as Europe and North America to watch the three days of competition. Most of them were experts in the field of sheepdog trialling.

Andrew was almost as nervous as me. Together we'd spent the four hours of the morning's singles competition working out tactics. The course was proving a real test for those running just one dog. Virtually all the runs so far had chosen to outrun to the left. The two that had tried to run to the right had both got their dogs hopelessly lost. My worry was that I had no option but to run one dog in each direction and Swift would have to take this right-hand path.

A round of applause signalled that the last of the morning's single competitors had completed his run. Now it was over to the brace competitors, first a member of the Welsh team, then me, on behalf of England.

The public address system burst into life.

'And now we move on to the brace competition,' the announcer said. 'There will be a short delay while we make the course more difficult.'

I looked at Andrew in disbelief, my eyes nearly popping from my head. What I had already seen this morning looked plenty difficult enough.

Five minutes later the Welsh competitor was at the post. Andrew and I peered past the grandstands to try to see the new course, and how he was faring. The Welshman's rather frantic whistling was a pretty good indication that things weren't going quite to plan. I could just make out his right-hand dog disappearing around the wrong side of a piece of woodland that protruded into the field, at the top of the hill. It was obvious that this right-hand outrun was going to cause even more problems than it had for the single dogs this morning.

I bent down to the dogs and rubbed their noses, talking to them quietly. Greg as ever looked happy and relaxed, wagging his tail at a passing dog. Swift was already intent on trying to sneak a look at the sheep out on the course.

There was one other friendly face as I endured the long wait. John Thomas, having won the singles event at the English National, was the team captain for the year, and now stood at the gate to the course, offering encouragement to the English competitors.

'When your right-hand dog goes over that bank give her a blast on the whistle,' he said, referring to the first obstacle, a small hill that Swift would have to go up and down before even starting the climb up towards the sheep, and the potential confusion of the woods above. John had forty years' experience of competition, and was a former supreme champion. He could read this course like a book.

John's words rang in my ears as I made the short walk to the post. I tried not to look behind, but I could almost feel the weight of the eyes watching me. Half a mile away on the side of the hill, ten sheep made their way from the letting-out pen. For a moment it seemed inconceivable, almost ridiculous, that the two dogs at my side were going to miraculously find their way across a Welsh hillside to collect sheep, with no more than a word of a command from me. Fortunately, the dogs didn't seem to share my sudden lack of confidence.

Greg had switched from being laid-back and distracted, to being focused and trembling with anticipation. He had clearly seen his quarry. Swift was more of a concern. She was looking in the right direction, but it was obvious from the look on her face that she had no idea where she needed to head. I gave Swift a quiet right-hand whistle. She needed to be sent first, she had further to go. I left her for a few moments, then spoke quietly to Greg. 'Away, Greg.' He was gone before I got through the first syllable.

Swift was now climbing the first small hill, still looking unsure of herself. Her head was turned to the left looking for the sheep that were still some four hundred yards ahead. Soon she reached the top of the hill and disappeared over the other side out of view. All I could do now was wait. If she spotted the sheep, she would come back into my view in a few moments, not far from her destination. If she didn't, well, that didn't bear thinking about . . .

I checked on Greg. As I'd thought, he knew exactly where he was going, and was spot on course. Nothing was going to stop him.

Swift, however, still hadn't reappeared from behind the small hill. It was only then that I remembered what John had said as I'd walked on to the course. 'Give her a blast on the whistle as she goes over the hill.' In the heat of the moment I'd forgotten.

Time seemed to slow down. It seemed like an age that she'd been out of view. I didn't have any idea which way she'd gone, she could have turned left or right once over the hill, I had no means of telling. But I knew I had to give a command at this point. I guessed and blew a shrill hard right. She didn't appear, but a murmur from the audience in the upper rows of seating behind told me that they could see something.

I looked around – it was obvious that they could see something on the left of the hill. I guessed again and gave a command to turn Swift back up the hill behind her. If she was indeed where I guessed she was, she should soon be in view. If not . . . again I batted the thought away.

The relief I felt when she at last came into view moments later was indescribable. She started to climb the hill below the sheep. I knew that I could now guide her into position.

I checked again on Greg. While I'd been concentrating on Swift, he'd remained focused on the job in hand. He had now arrived behind the sheep at just the right point, but far more impressive was what he'd done on getting there.

As far as my score was concerned, I'd most likely lost most of Swift's points on the outrun. But if both dogs 'lifted' the sheep at the same time, then the points from here on wouldn't be affected. Greg was obviously aware that this was a brace run, and on arriving at the sheep had lain down and waited for a minute or so for his partner to arrive without a command from me. When Swift appeared, he simply got to his feet and moved across to his side of the sheep. I don't know how many people noticed his reaction, but it was something that I'll never forget.

Once the sheep were under way the rest happened in what seemed like a flash. I was now concentrating far too hard to be aware of the ranks of spectators. As dogs and sheep successfully negotiated three sets of gates, I felt a mixture of confidence and relief flowing back through me. After a little delay Swift shed the sheep into two groups of five. It wasn't until I approached the first and most difficult pen that I was suddenly aware of the crowd. A silence had descended over the stands, which now seemed closer than ever.

I held my breath as Swift inched the sheep towards the pen entrance. All five had chosen to stand and face her, a situation I always dreaded as it was likely that one would attempt to make a break past her. Eventually one turned away from her but, instead of looking into the pen, the sheep looked at me, obviously intent on trying to find a way past. I crouched low, and glared into her eyes. Neither dog nor shepherd may touch a sheep during the trial – to do so risks disqualification. There was nothing in the rules to stop me gritting my teeth and growling, however, so that's what I did. The sheep looked into the pen, and after a little more hesitation walked in, leading the rest of her party with her.

'Good girl, Swift,' I said, stroking her head as she took up her position in the middle of the pen entrance. Now for Greg to pen his five sheep, always a precarious moment.

I decided to bring the sheep penwards in the slowest, most measured manner possible. If there were no sudden movements, I told myself, I'd just be able to nurture them inside the gate. But it wasn't to be.

I'd failed to position Greg in quite the right place, and in an instant the sheep bolted for freedom behind the pen. 'Half the points gone,' I said to myself. It was a minute later before Greg had them in position again. 'If Greg only pens once more in his life please let it be now,' I said to myself. My prayer was answered as suddenly all the fight seemed to evaporate from the quintet of sheep. With little further intervention from Greg, they wandered into the pen, leaving me to shut the gate behind them.

With applause ringing out from the stands, I called Greg and Swift to me and ruffled the collars of their thick coats. They had done all that I could have hoped of them.

John waited at the gate as I came off. Once more his simple 'Well done' spoke volumes. Andrew smiled, the relief showing on his face too. Whatever my points were, I had competed at the highest level in my chosen sport, and shown that the three of us were up to the challenge. Without Andrew I would never have been there. I would always be grateful for that.

With a further two days of competition left, we could now relax and enjoy things. Debbie, the children and my parents had all travelled up to watch me and the dogs. They'd sat in the stands during our run and were also quick to congratulate me. The children's appetite for sheepdog trialling was limited, however, and Debbie and my mother were soon whisking them off for some more child-friendly entertainment at the seaside a few miles away.

Andrew and I were leaning against the fence watching the remainder of the day's events when a familiar face appeared alongside us. He looked at us both, obviously trying to place where he'd seen us before.

'Didn't I sell you one of my pups a few years ago?' he said, in a familiar North Walian accent. It was Mr Jones who had bred Swift and we'd last seen at

Sennybridge, earlier in the year. 'How you getting on with her? Was taking a bit of handling, wasn't she, last time we spoke?'

I tried desperately not to sound smug.

'Not too bad,' I replied casually. 'Actually I ran her in the English team here earlier today.'

Mr Jones looked shocked. He reached for his programme and quickly thumbed through the pages until he spotted his own name, acknowledged as one of the breeders next to my entry.

'How'd she get on?' he asked excitedly.

By now, after their usual, hour-long deliberation the judges had put the day's scores up. With four judges on hand there were 560 points on offer. I'd scored 370 and was placed joint second.

'Not too bad, second at the moment, but there's another day tomorrow,' I said.

'Oh, well done, well done indeed,' he said, grinning from ear to ear.

I could have sworn there was an extra spring in his step as he walked off in the direction of the refreshments tent. I knew exactly how he felt.

CHAPTER TWENTY-NINE

The Turning of the Year

'Andrew, how are we ever going to find your Land Rover again?' It was a fair question, I thought. In a field, on the edge of Builth Wells, there must have been close to a thousand examples of Solihull's finest automotive export gathered around us – Land Rovers of every age, most attached to a livestock trailer.

The Builth Wells ram sale is the biggest of its kind in Europe, possibly the world. I hadn't been able to get to any of the ram sales in Devon over the previous month, so – with Andrew in tow – I'd made the by now familiar trip back up to mid-Wales. Andrew had been his usual positive self on the way. Even by Builth Wells standards that year's sale had attracted a huge number of sellers and there were almost ten thousand animals listed in the catalogue.

'They'll be cheap, bound to be, they'll never have enough buyers, you wait and see,' he said. I hoped he was right.

With each ram being responsible for seventy or more lambs each year, the rewards of spending time and money on choosing good stock are obvious. It still seemed a long way to come for three or four rams each.

The size of the auction was truly staggering. There were no permanent buildings at the Builth Wells showground – the whole sale was taking place in twenty huge marquees, in each of which was a rostrum, an auction ring and scores of pens containing rams brought in from across the country.

We entered one of the four tents selling Suffolks, and were hit immediately by the distinctive, overwhelming smell of ram, made all the more intense by the warmth of the canvas above. The rows between the pens of sheep were packed, thronging with farmers eyeing the next consignment to enter the ring. Unlike breeding ewes, rams are sold individually, and I always try to make a quick inspection of those on which I bid.

Buying rams can be a rather perilous occupation. First, there is a dazzling array of breeds, over twenty on offer today. The most popular by far is the Suffolk, which represented nearly a third of all the rams on offer, and the breed which I have always favoured for crossing with my Mule ewes. Rams are generally sold by their breeders at eighteen months of age and, in order to make the best possible price at this age, they need to be well grown, almost the bigger the better. But many breeders go to extraordinary lengths in order to get their rams to an appropriate size, feeding them high quality pellets up to five times a day. The result can be that when the magnificent looking specimens are taken home by the new owner and offered only modest rations, they quickly lose condition and after a few weeks look a shadow of their former selves.

The work that went into turning these rams out for sale never ceased to amaze me. In pen after pen owners worked puffing up wool, trimming backs and spraying fleeces. The results were truly magnificent to the eye, but rather misleading, which was why I liked to handle the sheep, to determine how much on show was wool and how much was the solid muscle that was actually needed.

I pushed past a couple of elderly men, leaning heavily on their sticks as they discussed the sheep in front of them, and climbed into the pen of enormous, jet-black-faced sheep, immediately to be accosted by the breeder.

'Late-born lambs these were,' he said in an accent that I assumed to be from Gloucestershire. 'Not born until April last year, take them home, sir, they'll grow like mushrooms.' Where have I heard that before? I thought, trying not to laugh. I spotted a trough full of cabbages in the corner of the pen. If they'd been fed on brassicas for the last eighteen months, they weren't likely to thrive at Mortehoe on a diet of rock and gorse. I made a hasty exit from the pen, pulled the catalogue from my pocket and put a cross through the number denoting these sheep.

An hour later and the sale was under way, and with it a miracle of modern logistics. In the twenty tents a dozen different companies of auctioneers started selling the ten thousand rams, reared by perhaps five hundred different breeders. But most incredible of all was the array of people to whom the rams were sold. Each auction tent was packed with farmers from across southern Britain, mostly completely unknown to the auctioneers to whom they made their bids, but by the end of the day the ten thousand rams would be loaded into the hundreds of livestock trailers in the car-park, and taken to their new homes. The money promised for each individual sheep would find its way from the thousands of bidders, via the dozen different auctioneers to the hundreds of sellers, and presumably everyone would be happy.

Andrew and I studied the catalogue as the Suffolks started to file through the ring. 'Won't bid on these, they've been on cabbages, nor the next pen, I looked in the mouths of several and the bottom jaws were overshot,' I muttered, with the catalogue raised to my mouth.

'I should give the next run a miss as well if I were you,' Andrew replied. 'I felt the back of one or two of them and there was nothing on 'em but wool.'

We had watched thirty or more rams pass through the ring before we saw anything that either of us thought suitable. I'd marked one particular ram down in my brochure, but no sooner had it entered the ring than the price had shot over the £400 mark, well over what I thought it prudent to splash out on this ram at least.

'It could be a long day,' I said to Andrew as we adjourned to the burger van, adding, 'I thought you said they'd be cheap here today.'

'They are, it's just you're being too fussy.'

Our patience was eventually rewarded, and between us we bought seven rams – four for me and three for Andrew. It was late afternoon before we collected them. With no hope of driving the Land Rover anywhere near the tent to load them, we resorted to marching the sheep individually from the tent to the trailer. It was no mean feat with each of the 200-pound rams hell bent on escaping into the sea of people thronging the alleyways. It was with some relief that we loaded the last one.

As we got ready to leave, the scene around us was a remarkable one. Everywhere, it seemed, sheep were appearing from marquees, ready to be loaded into Land Rovers, trailers, lorries and vans. I couldn't help thinking there was something unique about it. Where else would you find the entire contents of a sale, bought on the nod of a head and paid for with large, unguaranteed cheques? Somehow the market system still works for farming, as it has done for hundreds of years, with only minimal control. Long may it last, I thought to myself as we headed for home.

At Borough Farm, autumn had begun tightening its grip. Briefly that afternoon a shaft of sunlight had settled on the western-facing slopes of Borough Valley, lending the landscape a radiant, golden glow. But now as the clouds obscured the sun once more, a whipping wind was building, bringing with it a flurry of rustling leaves, blown from the cragged ash trees that stood on the exposed top of Steep Sheep.

As I made my way across the yard, the children were all out and about, wrapped up against the cold. Nick wore a woolly hat and gloves as he pulled his world-weary tractor and trailer through the sheep yards, and Clare and Laura appeared from the direction of the chicken runs in their winter coats.

'Where you off to, Dad?' Clare asked.

'To see the new rams. Coming?'

During their first night on the farm, the quartet of new rams had been integrated into the rest of the flock, penned tightly together with their new colleagues under cover. As the final week of September got under way, I'd already had cause to repeat the operation once after hearing the collision of skulls one evening. One of the new Suffolks was a lively specimen and I guessed I'd probably have to do it again before tupping started in a week or so.

Swift, once more, had been to the fore in dealing with the rams. Today I'd taken her along with Fern and Gail. Gail's progress over the summer had almost taken me by surprise. From the beginning, she had been remarkably easy to train, and perhaps because of this I had rather taken her development for granted. In her work she lacked the raw intensity that both Fern and Ernie possessed, but she had inherited from Greg a dependability that without my realising I was coming to rely on. It was she and Swift who drove the small flock of rams up towards me now.

The children hadn't had a chance to look over the new stock yet.

'Which one did you pay the most money for?' asked Clare, who seemed to be more and more aware of monetary matters these days.

'That one,' I said, pointing out the biggest of the quartet, for whom I'd paid £340.

'I don't like him,' Laura said. 'He looks too grumpy.'

A sudden chorus of high-pitched pip pips deflected the children's attention.

'What's that noise?' asked Laura.

'Swallows,' I said. 'Probably being chased by a sparrowhawk.'

A hundred or so swallows were darting across the darkening sky, pursued by the bigger, less agile sparrowhawk.

The hawk had no chance of catching his prey. The smaller birds were far too nimble, and when he did come close they ganged up in a mob to voice their disapproval.

The swallows were now gathering in flocks of increasing size and it wouldn't be long before they would leave on their long flight to Africa. Close to the farm fifty were already grouped together on a telephone wire, ready to head south. To me it was always a source of sadness to see them go. It would be a long, dark winter before they returned.

The cold westerly wind was growing in strength with each gust, whistling so loudly at times it drowned out the squabbling of the rooks returning to the valley at the end of their day. With the temperature dropping fast, Nick was beginning to look a little numb around the face so I picked him up and headed for home.

The weather forecast was predicting the first of the winter storms for later tonight and tomorrow, so I decided to head out to Morte Point with Gail before the light faded. I arrived there just after six, an hour or so before the official sunset but with the light already disappearing. In the deepening gloom, the peninsula felt deserted and somehow vulnerable.

The Morte flock tended to cause few problems at this time of the year and this evening I had little to do but turn a few ewes inland from the most exposed windward tip of the Point.

With a light whistle, Gail ran off into the howling wind, picking her way over the rocks before disappearing. When she reappeared she had three ewes scuttling along in front of her. I blew a long clear command, asking her to keep to her left so that the ewes took to a path leading inland. Amazingly, even with the roaring sea no more than a hundred yards behind, she heard the whistle and ran uphill, exactly as I wanted. This was Gail as I was coming to expect her, unspectacular, but highly effective and completely reliable.

I called her back to me and she returned, pushing her way through the bracken leaves already burnt by the sea salt and turning to a winter's brown once more. She wagged her tail, looking pleased with herself. I bent over and rubbed her nose, but she rolled on to her back to have her tummy tickled.

'You soppy dog,' I teased her as she lay there as if having a roll in the grass on a summer's day.

We walked a little way towards the very end of the Point, drawn once again by the compelling spectacle of the sea. By now the wind was tearing at my hair and hammering in my ears, making their drums ache from the pressure. Sea spray was beginning to wet my face, arriving in plumes from waves bursting on to the rocks a few yards below. The tranquil midsummer's paradise of only ten weeks ago had disappeared. If the eye of this storm was still twelve hours away, then tomorrow would be a wild day for sure.

Gail tucked herself close into my leg, her eyes half closed against the elements, her fur parted down one side by the strength of the wind. I turned my back on the wind and headed east. Ahead of me the Bull Point lighthouse flashed, its beam dulled by the dark grey rainclouds already driving in from the sea.

How little the physical landscape had changed since the lighthouse was first built in the late nineteenth century, when the wreckers had turned this section of the coast into 'the sailor's grave'. Yet how different were the lives of those of us who lived and worked within it.

As the last embers of daylight died away out in the Atlantic, the lights of the village glimmered dimly. Suddenly the thought of home, a warm kitchen and a piping hot cup of tea had an overwhelming appeal. With Gail at my side, I reached the sanctuary of the Land Rover just as the first, violent cloud burst began lashing at my back.

By the time I shut the cemetery gates a minute or two later, the entire peninsula had been consumed by a squalling, driving rain that blocked out the last traces of autumnal light. Winter – and a new shepherding year – would soon be upon us.

Acknowledgements

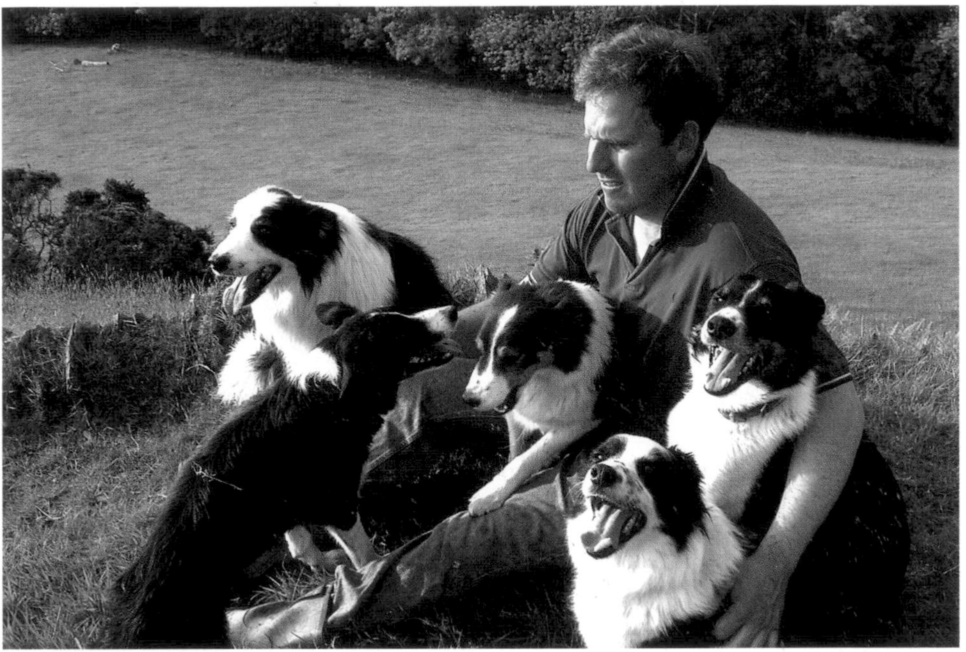

Becoming a published author may never have been the most likely career path for me, and I have many people to thank for this unexpected diversion from farming. My agent Mary Pachnos, for all her help, which includes everything from having the foresight to see the potential of the book, to putting me on the right bus when visiting London. To Val Hudson, Jo Roberts-Miller, Kerr MacRae and all the rest of the team at Headline, who inspire such confidence.

There have been many people along the way whose belief in a sheep-farmer with some 'novel' ideas was probably against their better judgement. Alan Parris, Steve Mulberry, Bill Baker, Jon Bourdillon and Beth Gregory to name but a few.

Guy Harrop's wonderful photography was achieved, not only through his understanding of the Devon countryside, but also through his willingness to lie face down in the mud, if the picture required. Thanks for additional pictures must also go to Joe Cornish, Brian Andrews, Ray Easterbrook and my daughter Clare who is responsible for the summer season opener.

Closer to home Derek and Helen, Garry and Cilene, and Andrew and Helen who in very different ways have been instrumental in the writing of this book.

I ought also to express my admiration for my sheepdogs past and present, who make my working life a pleasure and without whom I could claim to be neither author nor shepherd.

But I must save my biggest vote of gratitude for my family, Sarah, Martin, Mum and Dad, Clare, Laura, Nick and most of all Debbie. Thank you for all your support over some rather 'varied' recent years.